S0-BLS-274

ACS SYMPOSIUM SERIES 250

Ultrahigh Resolution Chromatography

Satinder Ahuja, EDITOR
CIBA-GEIGY Corporation

Based on a symposium sponsored by
the Division of Analytical Chemistry
at the 185th Meeting
of the American Chemical Society,
Seattle, Washington,
March 20–25, 1983

American Chemical Society, Washington, D.C. 1984

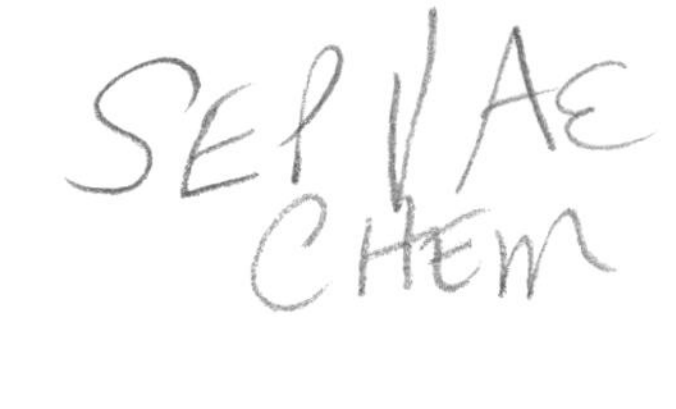

Library of Congress Cataloging in Publication Data

Ultrahigh resolution chromatography.
(ACS symposium series; 250)

"Based on a symposium sponsored by the Division of Analytical Chemistry at the 185th Meeting of the American Chemical Society, Seattle, Washington, March 20-25, 1983."

Includes bibliographies and indexes.

1. Chromatographic analysis—Congresses.

I. Ahuja, Satinder, 1933- . II. American Chemical Society. Division of Analytical Chemistry. III. American Chemical Society. Meeting (185th: 1983: Seattle, Wash.). IV. Series.

QD79.C4U48 1984 543'.089 84-2792
ISBN 0-8412-0835-2

PRINTED IN THE UNITED STATES OF AMERICA

ACS Symposium Series

M. Joan Comstock, *Series Editor*

FOREWORD

The ACS SYMPOSIUM SERIES was founded in 1974 to provide a medium for publishing symposia quickly in book form. The format of the Series parallels that of the continuing ADVANCES IN CHEMISTRY SERIES except that in order to save time the papers are not typeset but are reproduced as they are submitted by the authors in camera-ready form. Papers are reviewed under the supervision of the Editors with the assistance of the Series Advisory Board and are selected to maintain the integrity of the symposia; however, verbatim reproductions of previously published papers are not accepted. Both reviews and reports of research are acceptable since symposia may embrace both types of presentation.

CONTENTS

PREFACE

HIGH RESOLUTION CHROMATOGRAPHY evolved with capillary gas chromatography, a field that remained dormant for a long period of time because of the difficulty of preparation of suitable capillary columns. This difficulty has now been resolved, and an impetus has thus been provided for the development of ultrahigh resolution chromatography. Webster's dictionary defines ultra as beyond what is ordinary. Hence, in this book we will cover information that is beyond what is ordinarily considered high resolution chromatography—ultrahigh resolution chromatography. This information is at the cutting edge of the research on high resolution chromatography.

High pressure liquid chromatography, however, has made quantum jumps in the last decade and has provided several avenues for the development of ultrahigh resolution chromatography. Bonded columns with small particle sizes have revolutionized this field; a large number of theoretical plates are now possible with a 5-10 cm column with good resolution and reliability. Several coupled microbore columns or a long capillary column can provide an enormous number of theoretical plates and resolution that, a few years ago, were not considered possible. Microdetectors that provide minimal band broadening and optimum signal magnification have offered the necessary requisites for further development of this field. Mixed modes chromatography with or without column switching has made ultrahigh resolution a reality. Capillary GLC and HPLC are providing the main thrust in ultrahigh resolution chromatography today and are the main subjects covered in this volume.

Supercritical fluid chromatography (SFC) provides yet another approach to ultrahigh resolution chromatography and may provide resolution that is not now possible with GLC and HPLC. Because high solute diffusivity, lower viscosity, and excellent solvating properties can be obtained with supercritical fluids, high chromatographic efficiencies and fast analysis times can be obtained. Nonvolatile high molecular weight compounds can be separated at relatively low temperatures with efficiencies approaching those of GLC.

Ultrahigh resolution with other modes of separation still remains within the domain of a handful of experts. Although some information is included on these in the first chapter, we have deferred a detailed discussion of these techniques.

SATINDER AHUJA
CIBA-GEIGY Corporation
Suffern, New York

October 25, 1983

1

Overview: Multiple Pathways to Ultrahigh Resolution Chromatography

SATINDER AHUJA

Development Department, Pharmaceuticals Division, CIBA-GEIGY Corporation, Suffern, NY 10901

Today virtually every chromatographer understands what is implied by the term high resolution chromatography even though there is no hard and fast definition for it. Generally, high resolution gas chromatography entails gas-liquid chromatography with capillary columns and high resolution liquid chromatography involves the use of bonded phase packed columns with less than 30μ particle size. Webster's dictionary defines ultra as beyond what is ordinary. Hence we expect to cover in this book beyond what is ordinarily considered high resolution chromatography viz. ultrahigh resolution chromatography.

Even though there are no fixed rules to determine whether a given technique provides high resolution or not, the limits of resolution of any technique are well known to the practitioners. Hence, there is a constant thrust to improve resolution, i.e. separation between two or more components. Therefore, what constitutes high resolution is constantly changing. Since selectivity offered by each technique can be truly unique it would be desirable to arrive at some meaningful definitions in terms of N, α or k' values for high resolution in each technique. After this is accomplished, it would be easier to determine whether a given improvement leads to ultrahigh resolution.

Future symposia will address this problem. In the interim it is patently clear that, individually, none of the terms (N, α and k') in the resolution equation is sufficient to describe resolution (1):

$$\text{Resolution} = R_s = {}^{1}/_{4} \sqrt{N} \left[\frac{\alpha-1}{\alpha}\right]\left[\frac{k'}{1+k'}\right]$$

Where:

N = number of theoretical plates
α = separation factor
k'= peak capacity

It is erroneous to define resolution only in terms of N or k' as is frequently done by many chromatographers. An approach

0097-6156/84/0250-0001$06.00/0

that seems to have some merit is defining resolution in terms of the number of components that can be separated in a given unit of time. The time scale is a better measure of resolution than the commonly used theoretical plates scale. Of course, theoretical plates/second, attempts to address this question but misses the mark since the number of theoretical plates/second does not necessarily assure separation of a variety of components.

The frequency of component overlap is significant in chromatograms for which relative component spacing is governed by random factors. Giddings et al (2) calculate only 36% recovery for 100 components from a system with a peak capacity of 200. A 90% recovery would require a peak capacity of 1900 or 20 million theoretical plates. In those circumstances where a single component can be isolated as a single-component peak with an 80% probability, two components can be simultaneously isolated with a probability of only $(0.80)^2$ or 0.64. Thus the probability of isolating ten components simultaneously is quite small $(0.80)^{10}$ = 0.107). Consequently, with systems of enormous resolving power, it is difficult to resolve a small number of components simultaneously. This can have significant implications with respect to the way we handle data and design optimization procedures.

There are multiple pathways to achieve ultrahigh resolution because a variety of modes of separation are available. It is important to choose the best mode of separation for a given problem and optimize it rather than optimize a favorite mode of separation. Some of the modes of separation used today are: high performance thin-layer chromatography (HPTLC), capillary GLC, several modes of high performance liquid chromatography (HPLC), supercritical fluid chromatography, field flow fractionation, electrophoresis, electroosmosis and isoelectric focusing. Of these, capillary GLC and HPLC are providing the main thrust in ultrahigh resolution chromatography. Supercritical fluid chromatography is opening some new avenues and, hopefully, will provide resolution that is not possible now with GLC or HPLC. Hence, mainly these subjects have been covered in this book. Whereas, some improvements in resolution have been obtained with HPTLC, further improvements are possible in the area of detection and quantification. Ultrahigh resolution with the other modes of separation still remains within the domain of a handful of experts. Some information is included on these under Miscellaneous techniques, however, a detailed discussion of these techniques is deferred to a future symposium.

High Pressure Liquid Chromatography

A great deal of improvement is being made in general equipment to minimize band broadening. It is important to eliminate or minimize dead volumes. Special injectors with minimum dead volume are being experimented with to optimize separations. Similarly, detectors with small cell volumes are being developed to minimize post-column band broadening. As shown by the following equation, band-broadening can be minimized by giving consideration to each component:

$$W_T^2 = W_i^2 + W_f^2 + W_d^2 + W_c^2$$

injector fittings detector column

Major improvements are needed in conventional columns (4.6 mm i.d.) used for HPLC. Even though columns are available that can provide more than 10,000 plates/column or more than 100,000 plates/meter, large column to column variations can be found between manufacturers and from one batch to the next (3). Further improvements require better understanding of column variabilities and the separation processes occurring in the column.

An approach for improvement in this area has provided a new set of columns of smaller diameter, i.e. microbore columns. These columns can provide ultrahigh resolution when several columns are combined in a row. Due to the fact that a linear relationship can be obtained between efficiency and column length, columns of a million theoretical plates or more can be packed (4). Krejci et al (5) have described an open tubular column that is capable of providing up to 1,250,000 theoretical plates with an effectiveness of 50 theoretical plates/sec. The column (21 m x 60 μm I.D.) used 1,2,3-tris(2-cyanoethoxy)-propane as stationary phase and hexane saturated with stationary phase as mobile phase, with a linear velocity of 0.18 cm/sec. The above-mentioned plate count is was obtained with analysis time of 6 hours 54 minutes. Alternatively, a single column (330 μm diameter) packed with 3 μm C_{18} reverse phase packing can yield up to 110,000 plates for 1 meter length. Separation of priority pollutant mixture of 15 PAH components was demonstrated by Yang on one such column (6).

Ion-pair chromatography has provided numerous applications. Ultrahigh resolution can be obtained when a mixed mode is used i.e. one of the components is separated on the basis of ion-pair formation and the other is not (7).

Separation of Isophenindamine and phenindamine deserves special mention since an argentated HPLC mobile phase can bring about this difficult separation based on the position of double bond (8).

N-CH_3 N-CH_3

Phenindamine Isophenindamine

Several cholestric liquid crystals have been evaluated as stationary phases in HPLC (9). Cholesteric esters, for the most part, are liquid crystals in the range 20 to 100°C. Cholesteryl-2-ethylhexanoate was coated on Corasil II, and a mixture of these androgens was chromatographed between 0 and 80°C. The capacity factor increased 559% and 489% for androstenedione and $\Delta^{1,4}$ androstdienedione while the change for testosterone was ~100%.

Using a chiral recognition rationale, Pirkle, et al (10) designed a chiral fluoroalcoholic bonded stationary phase which separates the enantiomers of sulfoxides, lactones, and derivatives of alcohols, amines, amino acids, hydroxy acids, and mercaptans.

A cross-linked polystyrene resin with fixed ligands of the type (R)-N', N'-dibenzyl-1,2-propanediamine in the form of a copper (II) complex displays high enantioselectivity in ligand-exchange chromatography of amino acids (11). A microbore column (100 mm x 1 mm i.d.) packed with particles of dp = 5-10 μm generated up to 3500 theoretical plates, enabling a complete resolution of a mixture of three racemic amino acids into six components under isocratic conditions.

The combination of various chromatographic mechanisms (e.g. by the use of column switching) can expand the overall selectivity of the LC system by the nth power of that obtained with a single selectivity mechanism. Freeman (12) calculates a trillion compounds could be separated with the available column switching technology. Column switching can be viewed as a series of group separations where the individual compound separation is the last step i.e. final sub-classification into the sub-group.

Detectors such as laser fluorometry, FT-IR, mass spectrometer and flame-based detectors can be used to obtain desired selectivity and resolution. A phosphorus-selective detector for HPLC can provide ultrahigh resolution. In the phosphorus-selective chromatogram, the analyte peak is well resolved from neighboring peaks and the signal-to-noise ratio is also much higher than for the RI detector (13).

Capillary Gas Chromatography

High resolution chromatography evolved with the advent of capillary gas chromatography; however, this field remained dormant for a long period of time because of difficulties involved in the preparation of suitable capillary columns. Most of these problems have now been resolved providing an impetus for ultrahigh resolution chromatography.

One method of obtaining a very large number of theoretical plates by the exploitation of increased column length, avoiding the limitations imposed by the increased pressure drop, is that of recycle chromatography. Jennings et al (14) were able to generate in excess of 2,000,000 theoretical plates. It can be best employed where those components that can be separated by conventional high resolution GC are retained in that system, and an unresolved fraction is directed into the recycle unit to achieve separation. A route to "optimization" is based on maximizing the relative retentions of all the solutes in that sample (15, 16). This concept was employed to predict what lengths of dissimilar capillary columns should be coupled to achieve a specific binary liquid phase mixture column (17). Takeoka et al (18) suggested a method for the optimization of the separation of a model system, which could not be resolved on either of two 30 m columns coated with dissimilar liquid phases. One column required 25, and the other 57 minutes for the analysis; neither delivered separation of all components. Coupling calculated short lengths of the two columns permitted the complete separation of the mixture in less than 3 minutes.

Open-tubular columns utilized in gas chromatography have, with few exceptions, been of 0.2 mm diameter or larger and have therefore provided at most only a few thousand effective theoretical plates per meter length. Nevertheless, the theoretical background for design of column efficiencies of 10^5 or more effective plates per meter was presented as early as 1958. Laub et al (19) have explored the practical limits of the early theoretical work and in particular, the fabrication and properties of capillary columns of inner diameter ranging from 0.3 to 0.035 mm. The latter exhibits on the order of 2×10^4 N_{eff}/m for k' of 17. Much higher efficiencies could be realized with the mass spectrometric detector.

A comparison of efficiency of HPLC and GC columns was made by Widmer et al (20). For example, 3 μm packed 10 cm column gives 15,000 plates or 150,000 plates/meter. This compares with 11,880 plates obtained with 0.1 mm i.d. wall coated capillary GC columns. These data suggest that greater efficiency in terms of number of plates/meter is possible with HPLC.

Super Critical Fluid Chromatography (SFC)

Since high solute diffusivity, lower viscosity and excellent solvating properties can be obtained with supercritical fluids, higher chromatographic efficiencies and faster analysis time than liquid chromatography can be obtained with SFC (21). It is also possible to separate non-volatile high molecular weight compounds at relatively low temperatures.

While extremely large numbers of theoretical plates are possible with larger diameter columns (22, 23), calculations from chromatographic theory of the internal diameters and column lengths necessary to achieve relatively high efficiencies in reasonable analysis times indicate that column diameters of 50 to 100 μm i.d. are necessary for high-resolution SFC (23). For example, more than 10^5 effective theoretical plates are possible in less than two hours on 30-m long columns of 50 μm i.d.

UV and fluorescence detectors can be used for SFC. Carbon dioxide and nitrous oxide have made possible the use of conventional GC flame detectors (24). Based on studies of aliphatic fraction of a solvent refined coal product with supercritical CO_2 at 40°C and conventional FID detection, it was concluded (21) that the efficiencies of SFC are approaching those of GC (plate heights of 0.30 mm have been obtained in capillary SFC, which compare favorably with the 0.20-0.25 mm plate heights normally obtained in capillary GC).

Optimization scheme similar to that used for HPLC can be used for choosing a modifier for carbon dioxide based SFC (25).

Miscellaneous

A multichannel detector for in-situ analysis of fluorescent materials on TLC plates was investigated (26). The optical system was designed to obtain a fluorescence spectrum from each position along the elution axis of a one-dimensional plate without the mechanical scanning. By use of tetraphenylporphine and octaethylporphine it can be shown how overlapping spots can be resolved into their components.

Isomeric solutes that have closely related structure with slight difference in v.p. and degree of interaction with conventional stationary phases provide difficult separation problems. Liquid crystal, N,N'-bis(p-methoxybenzylidene)-α,α'-bi-p-toluidine, was used for separation of several PAH, in general, and benzo (a) pyrane and benzo (e) pyrene in particular (27). This stationary phase is mostly useful for isomers that are rigid to semirigid and differ slightly in their length-to-breadth ratio.

The related electrokinetic effects of electrophoresis and electroosmosis may be used to achieve ultrahigh resolution of charged substances (28). The electrophoresis is carried out in 75 μm i.d. glass capillaries. Reversed phase chromatography is performed by electroosmotic pumping of an acetonitrile mobile phase. The values of HETP for 9-methylanthracene and perylene are 19 μm and 25 μm respectively. The low reduced plate height values suggest column packing irregularities are less important when electroosmotic flow is used instead of flow generated by pressure.

Two dimensional electrophoresis provide ultrahigh resolution of various proteins (29). For example, by combining isoelectric focusing (IEF) with perpendicular electrophoresis in the presence of the detergent sodium dodecyl sulfate (SDS), the resulting resolution of 10,000 proteins could be theoretically obtained. Samples of human cells showed 100-2000 protein spots.

Literature Cited

1. S. Ahuja, Recent Developments in High Performance Liquid Chromatography, Metrochem '80, October 3, 1980
2. J. C. Giddings, J. M. Davis and M. R. Schure, This Symposium
3. S. Ahuja, Presented at New York Chromatography Society Meeting, June 10, 1981
4. R. P. W. Scott, J. Chromatog. Sci., 18, 49 (1980)
5. M. Krejci, K. Tesarik and J. Pajurek, J. Chromatog., 191, 17 (1980)
6. F. J. Yang, This Symposium
7. S. Ahuja, P. Liu and J. Smith, Personal Communication, August 7, 1979
8. R. J. Tscherne and H. Umagat, J. Pharm. Sci., 69, 342 (1980)
9. P. Taylor and P. L. Sherman, J. Liq. Chromatog., 3, 21 (1980)
10. W. H. Pirkle, D. W. House and J. M. Finn, J. Chromatog., 192, 143 (1980)
11. V. A. Davankov and A. A. Kurganov, Chromatographia, 13, 339 (1980)
12. D. Freeman, This Symposium
13. T. L. Chester, Anal. Chem., 52, 1621 (1980)
14. W. Jennings, This Symposium
15. H. J. Maier and O. C. Karpathy, J. Chromatog., 8, 308 (1962)
16. R. J. Laub and J. H. Purnell, Anal. Chem., 48, 799 (1976)
17. D. F. Ingraham, C. F. Shoemaker and W. J. Jennings, J. Chromatog., 239, 39 (1982)
18. G. Takeoka, H. M. Richard, M. Mehran, and W. Jennings, Personal Commuication
19. B. J. Lambert, R. J. Laub, W. L. Roberts and C. A. Smith, This Symposium
20. H. M. Widmer and K. Grolimund, This Symposium
21. W. P. Jackson, B. E. Richter, J. C. Fjeldsted, R. C. Kong and M. L. Lee, This Symposium
22. S. R. Springston and M. Novotny, Chromatographia, 14, 679 (1981)
23. P. A. Peaden and M. L. Lee, J. Chromatog., 259, 1 (1983)
24. J. C. Fjeldsted, R. C. Kong and M. L. Lee, Personal Communication
25. L. G. Randall, See This Book
26. M. L. Gianelli, J. B. Callis, N. H. Anderson and G. D. Christian, Anal. Chem, 53, 1357 (1981)
27. W. L. Zielinski, Jr., Indus. Res./Develop., p. 178, Feb. 1980
28. J. W. Jorgenson and K. D. Lukacs, J. Chromatog., 218, 209 (1981)
29. N. L. Anderson, Trends Anal. Chem., 1, 131 (1982)

RECEIVED November 10, 1983

Test of the Statistical Model of Component Overlap by Computer-Generated Chromatograms

J. CALVIN GIDDINGS, JOE M. DAVIS, and MARK R. SCHURE

Department of Chemistry, University of Utah, Salt Lake City, UT 84112

An earlier statistical model of component overlap is reviewed. It is argued that a statistical treatment is necessary to reasonably interpret and optimize high resolution chromatograms having substantial overlap.

A series of computer-simulated chromatograms has been generated to test the validity of a procedure derived from the statistical model for calculating the number of randomly distributed components when many of them are obscured by overlap. Plots of the logarithm of the peak count *versus* reciprocal peak capacity are used for this purpose. These plots are shown to provide reasonable estimates of the total number of components in the synthetic chromatograms.

In recent work, two of the present authors developed a statistical theory describing the frequency of component overlap in chromatograms for which relative component spacing is governed by random factors (1). It was our conclusion that the statistical theory would be most relevant for high resolution chromatographic systems, particularly when applied to complex mixtures. In fact, the statistical theory might be quite essential in dealing in any realistic way with exceedingly complex mixtures. For these samples, total resolution by chromatography is still beyond convenient realization.

The primary conclusion of the statistical model is that there is an unexpectedly high overlap of single-component peaks in chromatograms having randomly spaced components (1). Some examples best illustrate this point. In the first example, we considered a moderately high resolution system capable of resolving 100 peaks to some desired level. This is equivalent to saying that the peak capacity n_c is equal to 100. If one attempts to resolve 50 components on such a column - thus utilizing only half of the column capacity - it is found that only 18 of the 50 components appear as single-component peaks. Of the remaining components, roughly 14 appear in double peaks (a peak consisting of two overlapping components), 9 in triple peaks, 4 in a

0097-6156/84/0250-0009$06.00/0

quadruplet peak, and 5 components appear, on the average, in a quintet or larger peak. The same proportions exist for chromatographic systems of higher resolution. Thus an attempt to apply a system with a peak capacity of 200 to a 100-component mixture would lead to the isolation of only 36 components in a suitable way, that is, as single-component peaks. This represents a very unsatisfactory 36% recovery of desired information regarding the 100-component mixture.

It was also shown in the previously cited work that to increase the 36% recovery of single-component peaks to a 90% recovery, the peak capacity must increase from two to a factor of 19 times larger than the number of components. Thus to isolate 90 components from a 100-component mixture would require a chromatographic system with a peak capacity of 1900. This is equivalent to about 20 million theoretical plates, a system of such high resolution that it is beyond present reach. Unfortunately, only 5.3% of the theoretical peak capacity of such a system would be utilized by the 100 components due to the random spacing.

The difficulties posed by the above example are multiplied if one must isolate a number of components simultaneously from the chromatogram. This increase in difficulty stems directly from the approximate multiplicative nature of the probabilities involved. In those circumstances where a single component can be isolated as a single-component peak with, say, an 80% probability, two components can be simultaneously isolated with a probability of only $(0.80)^2$ or 0.64, and so on. Thus the probability of isolating a moderate number of components simultaneously, say 10, is quite small, e.g. $(0.80)^{10} = 0.107$. Consequently, even in systems of enormous resolving power, it is difficult to get, with any certainty, a small number of components simultaneously isolated.

The above difficulties have always existed with complex mixtures, but they have not always been clearly recognized. Often, we believe, chromatograms are interpreted too optimistically, without due regard for overlap, leading to mistakes in interpretation. The statistical viewpoint has brought this problem into much clearer perspective. Statistical models can now undoubtedly help guide researchers past the obstacles and lead them to more solidly based conclusions. Future statistical work should be directed at a broad spectrum of topics which can enhance the quality of chromatographic data, provide procedural guidelines, and firmly implant warning signals for important pitfalls involving overlap.

By way of example, it is clear that column switching, recycling, and optimization procedures can improve the pessimistic situation noted above. However, for complex mixtures, these procedures are also subject to statistical considerations. One of the objects of future work, therefore, is to evaluate the impact of statistical factors on general optimization procedures.

While the basic foundations of the statistical theory have been established (1), work on the above goals is still in its early stages. The main purpose of this paper is to test some of our statistical assumptions and procedures on various chromatograms. To this end, we have developed and analyzed 240 complex computer-generated chromatograms. We have chosen computer simulation because it would require an extremely elaborate experimental program to obtain this number of chromatograms using mixtures containing a known number (ranging up to 240) of components. In the end, if we found deviations from theory, we would not know if the deviations were due to shortcomings in the procedure or due to some departure from the statistical hypothesis. By using computer-generated chromatograms, we have total control over the statistical nature of the emerging "components". For example, we can precisely control noise, tailing, amplitude variations, baseline drift, and other important parameters that may have some effect on the numerical results.

The statistical theory we have developed is based on the hypothesis that component peaks are spaced randomly along part or all of the elution axis. We do not insist that random spacing is universal; in fact many chromatograms display obvious order. However, disorder appears to govern regions of other chromatograms, particularly those involving the ultra-high resolution chromatography of complex materials. By pursuing a theoretical model and attendant procedures based on the random spacing concept, we feel that criteria for judging randomness versus order in practical chromatography will emerge. The outcome of the chaos versus order question, which has not been seriously addressed before in chromatography, will have important implications with respect to the way we handle data and how we design column switching operations and other optimization procedures. At a more fundamental level, this work is designed to sharpen the procedures necessary to focus on factors which actually determine the structure of chromatograms and, by implication, the general statistical structure of complex mixtures that nature routinely puts into our laboratory, our resource base, and our environment.

Theoretical Background

Our treatment is based on the assumption that single-component peaks distribute themselves along separation axis x (e.g., elution time) according to Poisson statistics. This simply means that as we proceed along x, the probability that the component center of gravity falls in any small interval dx is given by $\lambda\, dx$, where λ is a constant or, at most, varies slowly with x. This assumption leads to the following probability distribution for the number of components m to be found in coordinate space X

$$P(m) = \frac{(\lambda X)^m e^{-\lambda X}}{m!}, \quad m = 0,1,2,\ldots \qquad (1)$$

The mean or expected number of components to be found in interval X is given by

$$\bar{m} = \lambda X \qquad (2)$$

The distribution in the distances between successive component centers is given by

$$P(x) = \lambda e^{-\lambda x} \qquad (3)$$
$$0 \leq x < \infty$$

and the probability that this distance falls below a critical value x_0 required for successful resolution is

$$P(x < x_0) = 1 - e^{-\lambda x_0} \qquad (4)$$

The critical distance x_0 is related to the critical resolution $R_s^{\neq}$ by

$$x_0 = 4\sigma R_s^{\neq} \qquad (5)$$

where σ is the mean standard deviation of the two adjacent single-component peaks. The dimension of each of the quantities, x_0 and σ, is determined relative to the generalized coordinate axis x which may represent length, volume, or time.

Pure components will elute from a chromatographic system in the form of underlying component peaks, roughly Gaussian in shape. However, the underlying component peaks often overlap significantly. One then observes the envelope of the merged component peaks rather than the individual component peaks. To clearly distinguish the difference between the peaks observed in chromatograms and the underlying component peaks, we define an <u>observed peak</u> or simply a <u>peak</u> as any envelope of a cluster of one <u>or more closely bound component</u> peaks. By closely bound, we mean that each underlying component peak is bound into the cluster by virtue of having a resolution of less than the critical value $R_s^{\neq}$ with respect to at least one other underlying component peak in the cluster. If the underlying component peak has a resolution greater than $R_s^{\neq}$ with respect to all other underlying peaks, that underlying component peak is considered to be an observed peak all by itself; it is termed a single-component peak or singlet.

We now note that peak capacity n_c is defined as being the maximum number of observed peaks which can theoretically be packed into a chromatogram or a specified portion of a chromatogram. Since the minimum observed peak spacing, according to the above definitions, is $x_0 = 4\sigma R_s^{\neq}$, peak capacity refers to the number of peaks which can be packed into the chromatogram with a spacing between adjacent peaks of x_0. When x_0 is approximately constant, as in many programmed separations, we have (1)

$$n_c = (t_2 - t_1)/x_o = (t_2 - t_1)/4\sigma R_s^{\neq} \quad (6)$$

which applies to the elution time range from t_1 to t_2. (When separation coordinate x is expressed in time units, then x_o and σ must, of course, also be expressed in time units.) In the absence of programming, we have

$$n_c = \frac{N^{1/2}}{4R_s^{\neq}} \ln \frac{t_2}{t_1} \quad (7)$$

where N is the number of theoretical plates.

The statistical theory shows that the mean number p of observed peaks is related to the expected number of components $\bar{m}$ by

$$p = \bar{m}\, e^{-\alpha} \quad (8)$$

where

$$\alpha = \bar{m}/n_c \quad (9)$$

Since α is the ratio of the mean component number $\bar{m}$ to the system's theoretical capacity n_c for separating peaks, α can be considered as a saturation factor for the chromatogram.

The statistical theory also shows that the number of one-component or singlet peaks (an observed peak consisting of only one underlying component peak) is given by

$$s = \bar{m}\, e^{-2\alpha} \quad (10)$$

Since it is the normal objective of chromatography to isolate individual components from all other components with a resolution of at least $R_s^{\neq}$, the number of singlet peaks relative to the number of components $\bar{m}$ is a measure of the success of separation. Unfortunately, Equation 10 shows that the $s/\bar{m}$ ratio falls off rapidly with increase in saturation factor α.

Experimentally, one can only see and count the observed peaks in a chromatogram; there is no way to tell whether those peaks are singlets or not. While one can count p, representing the number of peaks, the number of components $\bar{m}$ is the quantity of intrinsic interest. We have suggested that $\bar{m}$ is obtainable by writing Equation 8 in logarithmic form (1)

$$\ln p = \ln \bar{m} - \bar{m}/n_c \quad (11)$$

where Equation 9 has been used for α. Peak capacity n_c can be altered by variations in column efficiency. A plot of ln p *versus* $1/n_c$ should then yield $\bar{m}$ either by its relationship to the intercept ($\ln \bar{m}$) or the slope ($-\bar{m}$).

One of the major objectives of this paper is to test the above procedure for obtaining $\bar{m}$ using computer-generated chromatograms.

Computational Procedure

For the production of Gaussian-component chromatograms, we used a sum of individual Gaussians of the form

$$h(t) = B + \sum_{i=1}^{m} A_i \exp\left[-\frac{(t-t_{ri})^2}{2\sigma_i^2}\right] \quad (12)$$

In Equation 12, $h(t)$ is the amplitude expressed as a function of time t. The constant A_i is the peak height at t_{ri}, the retention time of the ith component peak, and σ_i is the standard deviation of the ith peak. The constants B and m are, respectively, the baseline offset and the total number of component peaks. The baseline offset may be time dependent.

We use the variable m to represent the total number of components in the synthetic chromatogram instead of $\bar{m}$. The former value is known in our computer-generated chromatograms but not in a complex mixture subjected to chromatography. In either case, only the mean component number $\bar{m}$ may be estimated by the statistical theory.

The units used for the amplitudes $h(t)$, A_i, and B are in analog to digital converter (ADC) units, which span from zero to 4095, corresponding to a 12 bit ADC.

To simulate chromatograms realistically and in a most general manner, the amplitudes of the Gaussians were assumed to be random. The amplitude A_i of the ith peak was calculated as

$$A_i = (A_{max} - A_{min})\chi + A_{min} \quad (13)$$

where A_{max} and A_{min} are constants corresponding to the maximum and minimum amplitudes, and χ is a random number between 0 and 1 having a uniform distribution. The chromatograms were also based on a uniform distribution of retention times between the limits of t_{min} and t_{max} such that

$$t_{ri} = (t_{max} - t_{min})\chi + t_{min} \quad (14)$$

The standard deviation, σ_i, in Equation 12 was constant in the calculation of the amplitude $h(t)$. This approximates the case of programmed temperature gas chromatography and other programming and gradient techniques.

Technically, the above procedure does not yield a Poisson distribution because m rather than λ is fixed. However, the approximation should be very good for large m.

In studies where it was desired to simulate noise, two forms of noise addition could be used. These are uniform and Gaussian noise (2); we chose the latter since short-term noise in chromatograms is usually Gaussian. For uniform noise, the time-dependent baseline amplitude B(t) is given as

$$B(t) = B + (1/2 - \chi)n_a \qquad (15)$$

where n_a is the noise amplitude. For Gaussian noise, the polar method (2) was used to produce Gaussian-distributed random numbers such that

$$B(t) = B + \chi_g \qquad (16)$$

where χ_g is a Gaussian deviate of standard deviation n_a and zero mean.

The computer program used here is written in FORTRAN and run on a PDP-11/34 minicomputer equipped with a 70 megabyte hard disk and a Versatec plotter. The operating system is RSX-11/M. Input data is held in files permitting unattended operation. This computer program, written in-house, is available from the authors upon request.

In many circumstances, we desired to change the random sequence of retention times and amplitudes produced by the random number generator. This was accomplished by allocating two unique numbers at the start of a program run as the random number seeds. The random sequence is unique for each different initial seed input. The random number generation of noise is programmed to be independent of the generation of retention times and peak amplitudes.

Two computer-generated chromatograms are shown in Figure 1. Both are simulations of the random distribution of 160 Gaussian components, each of which has a standard deviation σ of eight seconds, over a total separation space of 175 minutes. The amplitude range is 100:1800 and is scaled in terms of ADC units as described above. The total baseline peak capacity in both simulations is 219 (α = 0.731). The noise in the second simulation is Gaussian with a standard deviation of thirty ADC units.

A large number of overlapping peaks is visually discernible in these simulations, and yet the component number is considerably less than the peak capacity. Only 105 maxima are observed in the first and 99 maxima in the second simulation. Respective losses of 55 and 61 components result if each maximum is associated with a single component. These losses justify our earlier assertion that the total component number may be easily underestimated even from a high resolution chromatogram.

Methodology of Peak Counting

We restate the definition of a peak as a detected cluster of one or more components in which the first and last components in the cluster are separated from components adjacent to the cluster by a resolution greater than or equal to $R_s^{\neq}$ and in which each member of

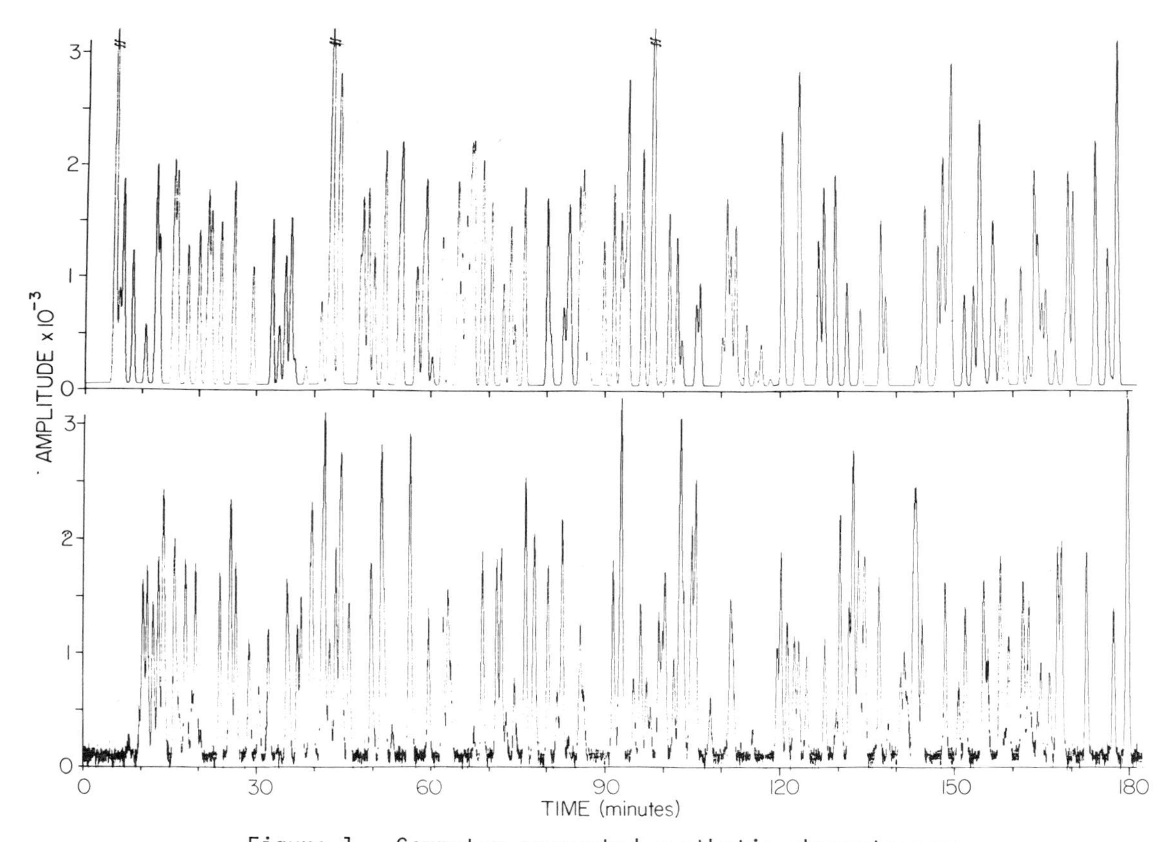

Figure 1. Computer-generated synthetic chromatograms. Above and below: 160 components, σ = 8 sec. The noise is Gaussian with a standard deviation of 50 ADC.

the cluster is separated from adjacent members of the cluster by a resolution of less than $R_s^{\neq}$. The number of peaks counted in a chromatogram is therefore dependent upon the critical resolution chosen. This result is implicitly incorporated into the model in terms of the column peak capacity which, in turn, is an explicit function of the critical resolution as shown in Equations 6 and 7.

The value chosen for resolution $R_s^{\neq}$ is dependent upon the requirements of the separation. If component detection is the sole objective, $R_s^{\neq}$ depends upon the methods available to discriminate between adjacent components. As an example, the sophisticated multi-wavelength spectroscopic detection of overlapping chromatographic components allows the magnitude of $R_s^{\neq}$ to approach very small values (3). Subtle differences in the mass spectra of eluting components can also be used to reduce $R_s^{\neq}$.

We recognize that many chromatographers, however, will interpret a chromatogram by visual inspection. This interpretation introduces difficulties because the resolution between two components is independent of the components' amplitude whereas visual differentiation between peaks is amplitude dependent. As an example, if one chooses to define each maximum in the chromatogram as a peak, the critical resolution between components of the same amplitude is 0.5 (4). If the adjacent amplitude ratio is 8:1, however, the critical resolution is approximately 0.8 (4).

One useful criterion is baseline separation. Peak counts based on the critical resolution for baseline separation, close to $R_s^{\neq} = 1.5$, are virtually independent of relative component amplitude. Peak counts are easily determined by counting as one peak all events between the rise from and decline to baseline. Several difficulties ensue, however, from this choice. First, this definition is not in accordance with the chromatographer's intuitive conception of a peak as a single maximum. Secondly, this choice of $R_s^{\neq}$ will give the smallest practical peak count at any peak capacity and thus the largest relative statistical error. The total component number is determined by plotting the logarithm of the peak count against reciprocal peak capacity (Equation 11). The error Δf in the function $f = \ln p$ is $\Delta p/p$ and will be large for small peak count p. (The error Δp will be dependent upon the method used to count peaks, upon noise, upon baseline stability, upon the statistical distribution of the peaks, etc.) Finally, baseline resolution is difficult to detect when there is baseline drift or noise.

A smaller value chosen for $R_s^{\neq}$ will increase the total peak count but will amplify the problem of relative component amplitudes. Because the model assumes no dependence of $R_s^{\neq}$ on relative component amplitude, we may, by one approach, use $R_s^{\neq}$ as a free parameter and choose its value so that the error between the simulation results and theory is minimized. The empirical value of $R_s^{\neq}$ is expected to depend on the relative component amplitude range, on the method of counting, and on the component density.

Once established, the value of $R_s^{\neq}$ may be used by chromatographers under the appropriate conditions.

Two methods of peak counting were used in this study. Both methods were based on visual inspection of the synthesized chromatograms. The criteria used to differentiate between peaks were baseline separation and resolution between maxima. The former was used to test directly the validity of the model independently of any empirical adjustment. The simulated chromatograms were synthesized with a flat and clearly discernible baseline before the first and after the final peak (see Figure 1). These segments were connected by straightedge and pencil. All events that lay between departure from and return to the baseline were counted as one peak. Our peak capacity calculations were based on $R_s^{\neq} = 1.5$ for the baseline separation case.

A clear and unambiguous maximum was the criterion used to determine the number of peaks using the second method. No peak shoulders were counted.

Gaussian noise was added to several simulations to determine its effect on both baseline and maxima peak counts. In the determination of baseline peaks, the center of the Gaussian noise was visually estimated, and a straight baseline was then sketched. The peak count was determined relative to this baseline as described above.

Simulation Studies

All of the results from the simulations described below are based on the assumption that component width σ is constant in a given simulation and that all components give Gaussian responses. The amplitudes of the components fell randomly within an eighteen-fold range. All chromatograms were generated over a 175 minute interval with five possible peak capacities as defined by component standard deviations σ of 12, 10, 8, 6, and 4 seconds and by some critical $R_s^{\neq}$. Three values of component number -- 80, 160, and 240 -- were chosen for the simulation studies.

A statistical analysis of the peak counts obtained from the simulated chromatograms was made as follows. We changed the random number sequence, by means of the seed change previously described, to generate random changes in component retention times and amplitudes while holding constant component number, zone width, and peak capacity. This procedure, in essence, mimicked the injection of different samples with the same component number and zone width onto a column. A mean peak count and standard deviation at each of the different peak capacities were calculated. The means and standard deviations of the peak counts were fit by a least squares analysis to Equation 11 with a proper transformation of the standard deviations from an exponential to a linear function (5). From the value of the least squares slope and intercept, an estimated component number was calculated.

Twelve simulation series, henceforth described as Set B, were generated for a noiseless baseline peak study with m = 160.

Each series, in turn, was composed of a unique distribution of components, which arose from the unique random number seeds, at each of five peak capacities (arising from the five different σ's).

Nine simulations at five different peak capacities were generated for each of two sets with Gaussian noise. The noise standard deviation was approximately one-third of the minimum component amplitude, A_{min}, in one series and one-half of the minimum component amplitude in the other. These simulation sets will be respectively described as Sets C and D. The component number m in each of these sets is also 160.

The empirical resolution $R_s^{\neq}$ from the counting of maxima was determined as follows. Maxima counts were made for all the simulation series, and averages and standard deviations of the maxima counts were calculated for each of the five peak capacities. A properly weighted least squares fit was made with the component number as a known quantity. The value of $R_s^{\neq}$ was then determined from the slope of the least squares fit.

This approach was possible because the function $\alpha = 4\sigma R_s^{\neq} \bar{m}/X$ contains the product $R_s^{\neq} \bar{m}$. The least squares analysis will give either quantity as the slope if the other is specified. This procedure allows the intercept to assume its optimal value instead of fixing it as $\ln(\bar{m})$. The latter approach or exact fit to theory, relative to the other method, underestimates $\bar{m}$ from the slope with virtually no difference in $\ln(\bar{m})$. This, in turn, arises because the maxima counting method over the range studied yields an intercept somewhat smaller than the theoretical one.

As previously stated, the empirical magnitude of $R_s^{\neq}$ should depend on component density. Two simulation series (Sets A and E, respectively) with component numbers 80 and 240 were also generated to determine the magnitude of this dependence. Each of these series was composed of nine noiseless simulations at five peak capacities.

Results and Discussions

Tables I and II are compilations of the total number of components estimated by using Equation 11 for the five data sets described above. Input data were the peak counts from a series of simulated chromatograms at five different n_c values for the three different component numbers. The results of Table I were obtained by a counting of baseline resolved peaks. Because these counts are nearly independent of component amplitude, these data are a rather direct test of the validity of the model and procedure. The results of Table II are, by contrast, based upon a counting of peak maxima and upon a calculation of an empirical resolution $R_s^{\neq}$. Table III contains data for a comparison of baseline peak counts from the simulated chromatograms to those predicted by theory.

The estimates of $\bar{m}$, from slope and intercept, calculated from

Table I. Simulation results for $\bar{m}$ using baseline counting. Amplitude range: 100-1800. Separation space: 175 min. Values of $\bar{m}$ are calculated from simulations at five different peak capacities based on peak standard deviations of 12, 10, 8, 6, and 4 sec. Noise is given in terms of amplitude units. Total peak capacity range: 146-438. S = $\bar{m}$ value from slope. I = $\bar{m}$ value obtained from intercept.

Component No. m, and Set	Noise	Number of Simulations	$\bar{m}$ From Weighted Least Squares (%Error) S	I	Mean & Std. Dev. of Data Sets S	I
80, A	0	9	82.3 (2.9)	78.7 (-1.7)	82.7± 14.7	79.0± 5.6
160, B	0	12	161.5 (1.0)	152.1 (-4.9)	164.2± 18.1	153.9± 8.9
160, C	30	9	145.6 (-9.0)	145.6 (-9.0)	147.7± 16.1	146.5± 9.2
160, D	50	9	149.0 (-6.9)	148.1 (-7.5)	153.9± 9.3	151.6± 6.3
240, E	0	9	266.1 (10.9)	245.9 (2.5)	267.3± 14.2	247.2± 18.6

Table II. Simulation results for $\bar{m}$ using maxima counting. Other conditions same as Table I except n_c not specified.

Component No. m, and Set	Noise	Number of Simulations	Least Squares Resolution, $R_s^{\neq}$	Mean & Std. Deviations of Data Sets S	I
80, A	0	9	0.499	84.5± 28.4	78.0± 3.2
160, B	0	12	0.515	159.4± 24.6	150.0± 7.8
160, C	30	9	0.496	155.1± 30.4	144.8± 5.2
160, D	50	9	0.427	159.9± 17.7	139.0± 6.8
240, E	0	9	0.482	241.4± 31.5	221.8± 11.1

Table III. Comparison of average simulation baseline peak counts and theoretical expectation. Amplitude range: 100:1800. Component standard deviations: 12,10,8,6, and 4 sec. Separation space: 175 minutes. Noise is given in terms of amplitude units. $\alpha = \bar{m}/n_c$. $R_s^{\neq} = 1.5$.

Component number and Set	Number of simulations	Noise	Component standard deviation, sec								
			12 (n_c = 146)			10 (n_c = 175)			8 (n_c = 219)		
			Simulation	Model	Alpha	Simulation	Model	Alpha	Simulation	Model	Alpha
80, A	9	0	44.1±2.9	46.2	0.55	50.0±3.7	50.6	0.46	54.6±3.6	55.5	0.37
160, B	12	0	53.1±4.1	53.4	1.10	62.7±3.8	64.1	0.91	75.7±2.7	77.0	0.73
160, C	9	30	51.2±3.8	53.4	1.10	63.6±2.4	64.1	0.91	76.2±4.6	77.0	0.73
160, D	9	50	50.2±4.8	53.4	1.10	59.9±4.7	64.1	0.91	72.4±4.6	77.0	0.73
240, E	9	0	39.2±2.4	46.2	1.65	54.2±2.3	60.9	1.37	72.3±4.5	80.1	1.10

Component number and Set	Number of simulations	Noise	6 (n_c = 292)			4 (n_c = 438)		
			Simulation	Model	Alpha	Simulation	Model	Alpha
80, A	9	0	59.1±3.7	60.8	0.27	65.0±3.0	66.6	0.18
160, B	12	0	88.0±4.6	92.4	0.55	104.0±4.1	111.0	0.37

the weighted least squares analysis are, on the whole, in agreement with the actual component numbers and confirm the general correctness of the results predicted by the model over the eighteen-fold amplitude range examined. One might initially attribute the underestimation of component number in Sets C and D to an inability to detect visually small peaks in the noisy baseline. However, comparisons of the total baseline peak count at each peak capacity for Sets B, C, and D by means of f- and t-tests reveal no statistical difference in the peak counts at a given peak capacity, with one exception, at the 95% confidence level, and no difference at the 99% confidence level.

Figure 2a is the weighted least squares plot of the logarithm of the number of baseline peaks counted in Set B versus reciprocal peak capacity. The error bar bracketing each point is the properly weighted standard deviation of the mean peak count at that peak capacity. The estimated component numbers from the slope and intercept are reported in Table I and are, respectively, 161.5 and 152.1. The close agreement between the estimated and actual component numbers verifies the reasonable estimation of component number from Equation 11.

Estimates of the component number were also made by an unweighted (weight of unity) least squares fit to the results from each of the nine or twelve simulation series composing the five sets A, B, C, D, and E. Each simulation series effectively mimicks an analytical mixture because the relative retention times and component amplitudes remain constant as peak capacity (column efficiency) changes. The results from an individual simulation series consequently are similar to those expected by an experimental variation of column efficiency for a given mixture. The results from each of the nine or twelve simulation series were averaged to give a mean value to compare against the least square analysis.

The average component numbers from these unweighted fits are in close agreement with the results from the weighted least squares analysis as expected. The standard deviation, and hence the unrealiability of the results from a single series, is generally higher in the estimation of component number from the slope than from the intercept. This observation is attributed to the variances in slope and intercept which, in turn, are functions of the peak capacities and the number of counted peaks (6). The variance in the estimation from the intercept increases with the value $\bar{m}$, and a reversal of the trend is observed in Set E.

Table II contains the optimal empirical resolutions calculated from the weighted least squares analysis of counted peak maxima. The empirical $R_s^{\neq}$ from each set was used to calculate, from an unweighted least squares fit, a component number from each series in that set. The results for each set were averaged, and the mean and standard deviation for each set are reported in Table II.

The overall optimal empirical resolution, over an eighteen-fold amplitude range, for the noiseless simulations is essentially independent of α (over a nine-fold range) and has the average value of $R_s^{\neq} = 0.498 \pm 0.017$, or approximately $R_s^{\neq} = 0.5$. In general, however, the optimal $R_s^{\neq}$ is expected to be a function of the total relative amplitude range. Studies are currently underway to determine $R_s^{\neq}$ for larger relative amplitude ranges.

The results from the individual series estimations were averaged for a comparison to the results from the weighted least squares analysis. The standard deviations in $\bar{m}$, as estimated from the slope, are generally larger from use of the maxima counting procedure than from the baseline counting procedure, whereas the standard deviations in the estimates from the intercept are comparable in both methodologies. This observation may again be attributed to the variances of the slope and intercept of the individual simulation series.

The value of the critical resolution $R_s^{\neq}$ estimated from the low noise power study, Set C, suggests small magnitudes of noise have little or no effect on the value of $R_s^{\neq}$. The estimation of $R_s^{\neq}$ from Set D, however, suggests large values of noise do exert an influence. This change arises because a smaller number of maxima are counted in the chromatograms with low signal-to-noise ratios than in the chromatograms with high signal-to-noise ratios. One possible origin of the low counts can be cited. Two peak maxima barely separated but distinguishable in a noiseless simulation are not resolved visually in a noisy simulation if noise on the two maxima masks the separation. They will be counted as one large peak and not two peaks. This hypothesis is under further investigation.

The critical $R_s^{\neq}$, as obtained from visual inspection, is a sensitive function of the noise level relative to the minimum peak amplitude. A systematic study would be required to determine a family of $R_s^{\neq}$ values as a function of the noise-to-minimum-peak-amplitude ratio. This limited study suggests caution in the application of the overlap model based upon maxima counts to noisy chromatograms. Small amounts of noise do not, however, appear to contraindicate the evaluation $R_s^{\neq} = 0.5$.

Figure 2b is a graph of the unweighted least squares results for counts obtained from three of the twelve different simulation series composing Set B. Baseline resolved peaks were counted. Each slope and intercept is a datum for the determination of the overall mean and standard deviation compiled in Table I for Set B. A large spread is noted in the estimations of the component number. The data points for each series appear to fall along a straight line, and no indication that results estimated from the upper and lower lines are in error can be gleaned from the data.

It may be argued that five data points are an insufficient source of statistical information. The following argument, however, is perhaps more relevant. The slopes and intercepts determined from the weighted least squares fit were calculated from a mean peak count determined from either nine or twelve simulations at each peak capacity. The average result is more in agreement with the model, which predicts only average results,

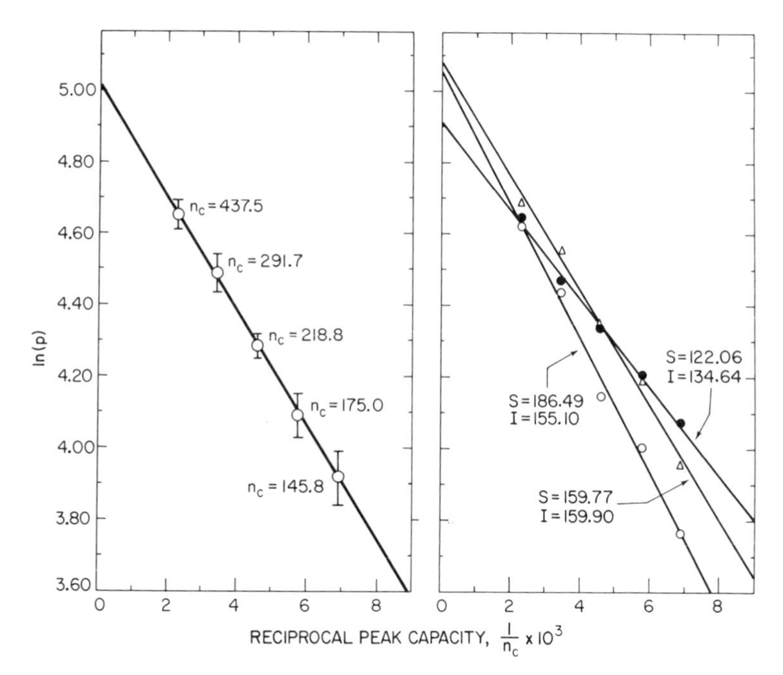

Figure 2. (a, left) Weighted least squares fit to Equation 11. m = 160. (b, right) Unweighted least squares fit to Equation 11 for for three simulation series. Numbers are the estimates $\bar{m}$ from slope (S) and intercept (I). m = 160.

than any single analysis. Accordingly, ways of randomizing the component retention times from a single sample are desirable to obtain a more accurate estimation of the component number. For a real sample, the component retention times may be slightly randomized by changing the mobile and/or stationary phase of the column. The changes in retention time from column to column will not be extensive because the physiochemical processes affecting retention (e.g., vapor pressure and solubility in GLC) will not change extensively, but the randomization may give rise to a closer estimation of the actual component number.

Table III contains the values of the actual mean baseline peak counts and the baseline peak counts predicted by the model. The full table is not presented due to space constraints. The agreement between the mean simulation counts and theory is fairly good when $\alpha < 1$ and confirms the validity of the model and procedure over the stated amplitude range. Frequently, a smaller number of peaks were counted than are predicted by theory when $\alpha > 1$.

The amplitudes of the individual components and constituent peaks create a departure from theoretical predictions in these highly saturated simulations when the peak number is visually interpreted. Consider a peak cluster composed of at least two overlapping components. The coordinates of the centers of gravity of the first and last components in the cluster are, respectively, x_1 and x_2. Therefore, the beginning of the first component in the cluster occurs along coordinate x at $x_1-x_0/2$, and the end of the final component in the cluster has coordinate $x_2+x_0/2$. The peak width is then postulated to be $x_2-x_1+x_0$ in our model.

This conclusion is based on the assumption that each component width is unaffected by other components that overlap with it and is correct if the zone width is represented as a one-dimensional vector, e.g., a line segment. However, a visual interpretation of a chromatogram cannot be made independently of amplitude. Any real peak has a finite amplitude, and amplitudes of two or more closely overlapping components will add constructively to extend the effective width of the overlapping cluster. The real peak is therefore wider than the theoretical peak, and a greater probability exists that another component will overlap with it than will overlap with the theoretical peak. The number of counted peaks consequently decreases below the number predicted by theory. (We note that this difficulty is based upon a visual interpretation of peaks and may not exist with another method of peak counting.) This departure is expected to be most severe when the number of components contributing to a peak is high ($\alpha > 1$).

This "third body amplitude" hypothesis is strengthened by a study that confirms the agreement between point statistics and the model even when $\alpha > 1$ and is under further investigation. Consequently, it is advisable that the model only be applied to a chromatogram in which the baseline $\alpha < 1$. The quantity $\alpha = \bar{m}/n_c$ is, however, *a priori* an unknown since the mean component number $\bar{m}$ is unknown. The means for its crude estimation from a chromatogram are desirable since the validity of the component number estimated from a series of visual peak counts is dependent on α. This problem is under current investigation.

Conclusion

The results of Tables I, II and III confirm the general applicability of the peak overlap model, developed from point statistics, to randomly generated chromatograms. Individual exceptions to the model will undoubtedly be found as experimental testing is conducted, but, overall, we anticipate modestly good predictions of $\bar{m}$ from high resolution chromatographic separations when the components are distributed randomly.

We extend an offer to interested experimentalists to test the underlying hypothesis of our model and the correctness of its predictions. A series of controlled studies using a mixture, either natural or synthetic, of a large known number of components with widely varying functional structure is one possible procedure. Only experimental testing will reveal the degree of applicability and universality of our model and its predictions.

Acknowledgment

This work was supported by National Science Grant No. CHE79-19879. One of us (JMD) acknowledges funding from the University of Utah Research Committee.

Literature Cited

1. Davis, J.M.; Giddings, J.C. Anal. Chem. 1983, 55, 418-424.
2. Knuth, D.E. "Semi-Numerical Algorithms"; The Art of Computer Programming, Vol II; Addison-Wesley: Reading, 1969; pp. 104-105, 113.
3. Kowalski, B. "Multivariate Curve Resolution in Liquid Chromatography"; 185th National American Chemical Society Meeting; Seattle, Washington, 1983.
4. Snyder, L.R.; Kirkland, J.J. "Introduction to Modern Liquid Chromatography"; John Wiley: New York, 1979; pp. 38-41.
5. Isenhour, T.L.; Jurs, P.C. "Introduction to Computer Programming for Chemists"; Allyn and Bacon: Boston, 1972; pp. 178-182.
6. Bevington, P.R. "Data Reduction and Error Analysis for the Physical Sciences"; McGraw-Hill: New York, 1969; pp. 115-116.

RECEIVED September 27, 1983

3

Ultrahigh Resolution Gas Chromatography

Constraints, Compromises, and Prospects

WALTER JENNINGS

Department of Food Science and Technology, University of California, Davis, CA 95616

Resolution...defined as the separation of components...is represented by

1] the degree to which the peak maxima are separated, and
2] the "sharpness" of the peaks.

The former is primarily affected by column contributions, which include

(a) column length;
(b) operational parameters such as the type of carrier gas and use conditions (e.g. velocity, temperature, pressure drop); and
(c) solute properties, including their partition ratios and relative retentions.

The "sharpness" of the solute peaks is affected by

1] column contributions (column length and the height equivalent to a theoretical plate);
2] operational parameters (e.g. carrier gas choice, conditions of use, temperature);
3] solute properties (e.g. partition ratios), and
4] the efficiency of the injection process.

For a given set of circumstances, resolution is maximized when

1] the band beginning the chromatographic process is as narrow as possible;
2] all the identical molecules of each solute behave in the same manner, i.e. the range of times they spend in the gas phase is as small as possible and the range of times they spend in the liquid phase is as small as possible; and
3] the interaction(s) between the molecules of each of the solutes to be resolved and the liquid phase are as different as possible.

The net effect of changing any one of these parameters is complicated by the fact that many of them are inter-related. In addition, in our quest for resolution we too often lose sight of the fact that analysis time is also an important analytical parameter, and many of these parameters we are considering also exercise effects on the analysis time.

0097-6156/84/0250-0027$06.00/0

A knowledge of the composition of the sample permits some intelligent decisions to be made, but in many cases, the composition of the sample has not yet been elucidated; the analyst doesn't know how many components are in it. One can only subject the sample to separation on the most efficient columns available, under conditions where those columns exhibit a maximum number of theoretical plates, and hope that each peak in the resulting chromatogram represents a separate solute, i.e. that resolution is complete. Here the lessons we learn from theoretical considerations can be put to practical use:

$$Rs = \frac{1}{4} \sqrt{n} \left(\frac{\alpha - 1}{\alpha}\right) \left(\frac{k}{k + 1}\right)$$

Obviously, anything that increases n, α, and/or k and that does not cause other deliterious effects will enhance resolution.

Columns of greater length and/or smaller diameter will yield more theoretical plates, but changing either parameter leads to a higher pressure drop through the column, and the van Deemter curves become sharper; because the van Deemter curve is k-specific, the range of solute partition ratios over which maximum resolution can be realized is limited (1). It is this latter difficulty that also limits the utilty of high pressure gas chromatography; some twenty years ago Desty demonstrated that by working at elevated pressures, many more theoretical plates could be obtained, but only over a very limited range of k-values (2).

One method of delivering very large numbers of theoretical plates by the exploitation of increased column length that avoids the limitations imposed by the increased pressure drop is that of recycle chromatography. Jennings et al. (3) were able to generate in excess of 2,000,000 theoretical plates; at u-opt, the recycle apparatus (Figure 1) delivered approximately 2000 n/sec, while at the OPGV, ca. 4000 n/sec were realized. The utility of this approach for general use is limited by the fact that only a restricted range of partition ratios can be accommodated within the system at one time. It was suggested (4) that it might best be employed in the configuration shown in Figure 2, where those components that can be separated by conventional high resolution GC are retained in that system, and an unresolved fraction could be directed into the recycle unit, cycled to achieve separation, and directed back into the mainstream of eluting solutes during a period of quiesence. The recent appearance of a new valve, designed to embrace zero dead volume when used with fused silica capillaries, which shows great promise of resolving many of our earlier difficulties, may eventually make possible the more routine use of recycle chromatography.

Lower temperatures (and lower program rates) will lead to increased values of k (and of α), and permit the system to deliver improved resolution, but another result will be longer analysis times, and (because the KDs are larger as the solutes are delivered from the end of the column to the detector,) lower sensitivities (1).

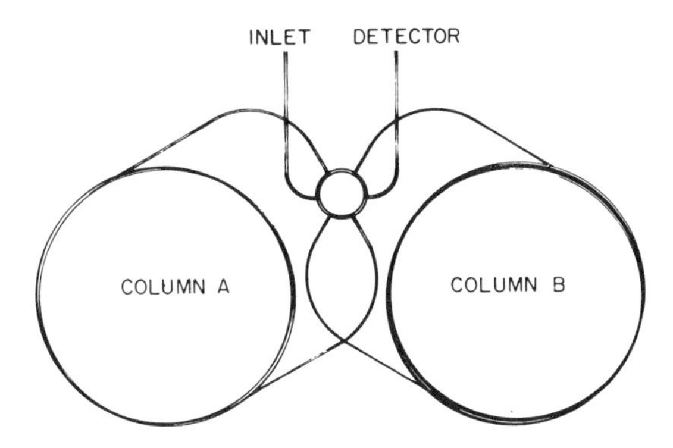

Figure 1. Schematic of a recycle unit on which theoretical plate numbers in excess of 2,000,000 were generated. The unit delivered ca. 2,000 n/sec at u-opt, and 4,000 n/sec at OPGV. From Jennings et al. (3). Copyright 1979, Elsevier Science Publishers B. V.

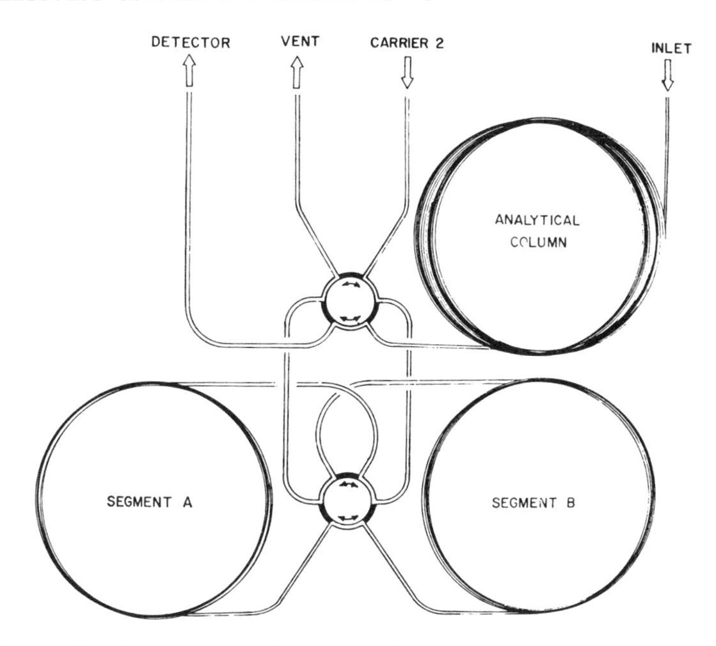

Figure 2. Schematic of a recycle unit in combination with a standard analytical column. A multi-component peak from the latter is directed into recycle where it is subjected to separation, and the resolved components re-directed back to the detector during a quiescent period in the normal analytical chromatogram. From Jennings et al. (4). Copyright 1979 Dr. Alfred Huethig.

Thicker films of liquid phase would also result in larger partition ratios, but as the film thickness increases beyond about 0.4 μm, we lose one of the major advantages of capillary gas chromatography: diffusivity in the liquid phase is no longer inconsequential. Figure 3 shows via computer plots (4) the predicted result of increasing d_f for an assumed set of conditions.

Relative retentions..the α values..usually vary inversely with column temperature, but are most strongly affected by the choice of liquid phase. In packed column chromatography, the choice of liquid phase is usually the most effective route by which separation efficiency is influenced. In capillary GC, however, there is normally such an abundance of theoretical plates that the choice of liquid phase is a relatively unimportant parameter for many analyses. In some cases however, it does become desirable (or even necessary) to select a liquid phase in which the relative retentions of certain solutes is larger. Until quite recently, this posed a real problem with the fused silica capillary column, because the more polar liquid phases, i.e. those in which relative retentions are usually greater, coated fused silica only reluctantly, and produced columns whose useful lives were quite limited. The development of stable bonded phase columns (5) eventually overcame this difficulty (vide infra).

It is possible to exploit liquid phase selection to an even greater degree: if we know what solutes are in the sample, we can select a "liquid phase" that maximizes the relative retentions of all the solutes in that sample. This route to "optimization" was first suggested by Maier and Karpathy (6), and explored by Laub and Purnell (7). It was these latter authors who gave us the concept of "window diagrams". Ingraham et al. (8) employed these concepts to predict what lengths of dissimilar capillary columns should be coupled to achieve a specific binary liquid phase mixture, and discussed the complexities that were introduced by the gradient in carrier gas velocity that exists through the column, which produces an "apparent" liquid phase ratio differing from that actually present. Takeoka et al. (9) suggested a method for determining the "apparent" liquid phase ratio, and making the necessary corrections so that it coincided with the desired phase ratio. This permitted optimization of the separation of a model system, which could not be resolved on either of two 30 m columns coated with dissimilar liquid phases. One column required 25, and the other 57 minutes for the analysis; neither delivered separation of all components. Coupling calculated short lengths of the two columns permitted the complete separation of the mixture in less than 3 minutes.

From the relationship

$$h_{min} = r\sqrt{\frac{11k^2 + 6k + 1}{3(1 + k)^2}}$$

it can be deduced that at equivalent coating efficiencies n varies inversely with r; this has led to some recent interest in very small diameter columns (e.g. (10)). However, such columns are more easily overloaded, and impose more stringent demands on injection hardware and its use. Also, their pressure drop per unit length is significantly higher, leading to sharper van Deemter curves. It is probably because of these considerations that such columns are usually limited to short lengths which can be exploited to yield equivalent separations in much shorter times.

One of the developments that promises to greatly improve our prospects for increased resolution is the very recent availability of high quality fused silica columns, with quite an array of more polar, highly cross-linked liquid phases. Figures 4-8 show chromatographic separations obtained on several such columns. There is promise of continued progress in this area, which, in combination with the other routes to improved resolution considered above, must ultimately result in far superior chromatographic separations.

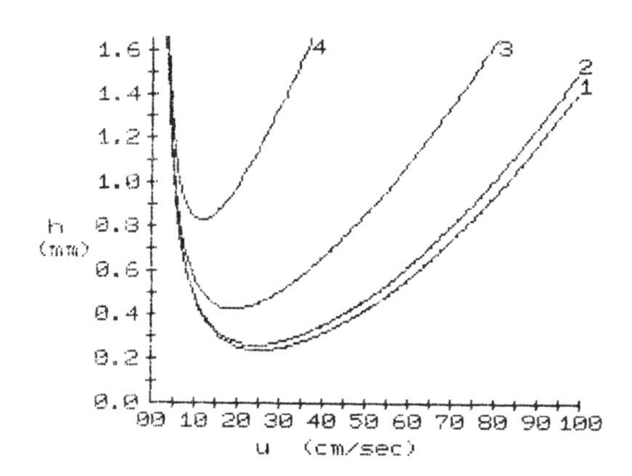

Figure 3. Theoretical van Deemter curves, generated from a computer program presented by Ingraham et al. (5), for 0.25 mm x 30 m columns with helium carrier. Partition ratio of the test solute, k = 5; D_l taken as 10^{-6} cm^2/sec, D_g as 0.3 cm^2/sec. Curve 1, d_f = 0.25 μm; curve 2, d_f = 0.4 μm; curve 3, d_f = 1.0 μm; curve 4, d_f = 2.0 μm.

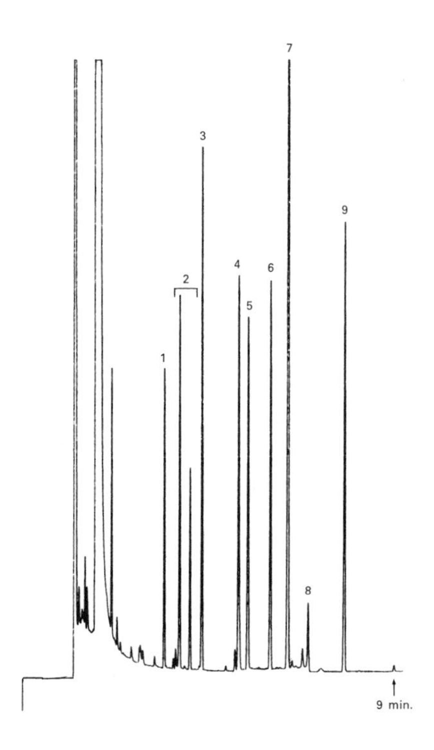

Figure 4. TMS derivatives of several sugars as resolved on the 7% phenyl-7%-cyanopropyl-polymethyl siloxane liquid phase DB-1701. Split injection (in acetonitrile/TMS imidazole 1:1) on a 30 m x 0.25 mm column coated with a bonded 0.25 µm film. Hydrogen carrier at 40 cm/sec; 180°C 2 min, and 5°/min to 200. Components: 1, D-arabinose; 2, D-ribose anomers; 3, D-xylose; 4, D-mannose; 5, D-fructose; 6, α-D-galactose; 7, α-D-glucose; 8, D-fructose anomer(?); 9, β-D-glucose.

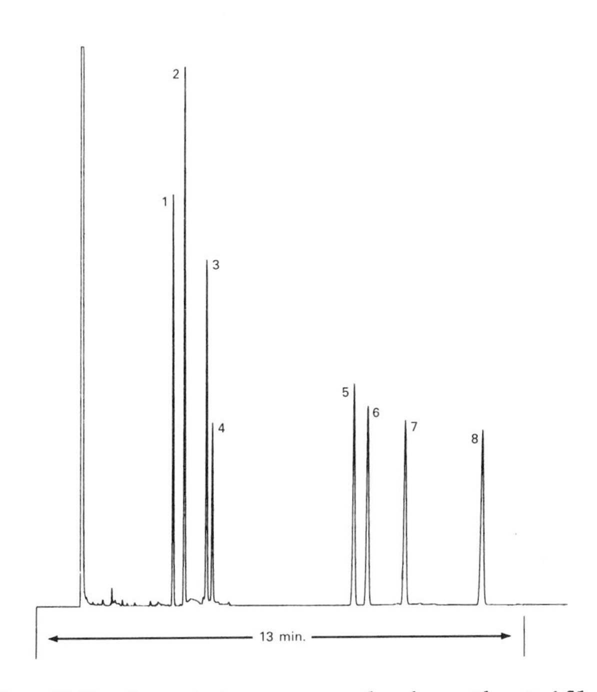

Figure 5. Alditol acetates as resolved on the trifluoropropylmethyl silicone DB-210. Split injection, isothermal at 220°C; 30 m x 0.25 mm column coated with a bonded 0.25 µm film. Components: 1, rhamnitol; 2, fucitol; 3, ribitol; 4, arabinitol; 5, mannitol; 6, galacitol; 7, glucitol; 8, inositol.

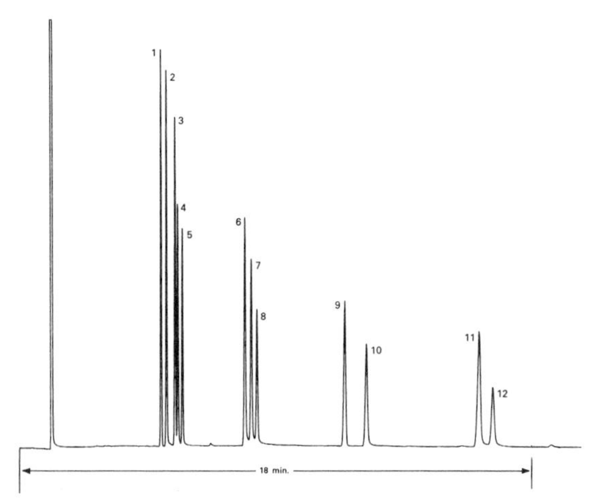

Figure 6. Free underivatized steroids as resolved on the 50% phenyl-polymethylsiloxane DB-17. Split injection at 260 °C isothermal; 30m x 0.25mm column coated with a 0.15 m bonded film of DB-17, hydrogen carrier at 44cm/sec. Components: 1, coprostane (5-B-cholestane); 2, 5-β-androsterone; 3, 5-α-cholestane; 4, androsterone; 5, epiandrosterone (trans-androsterone); 6, 17-estradiol; 7, -estradiol; 8, estrone; 9, progesterone; 10, cholesterol; 11, estriol; 12, stigmasterol.

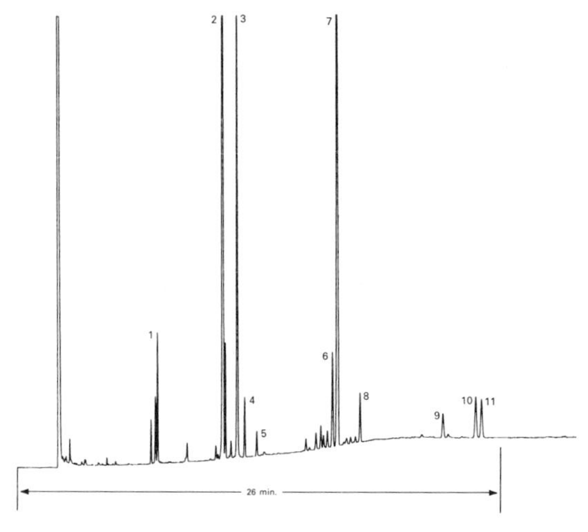

Figure 7. Methyl esters of unsaturated fatty acids as resolved on the 25%-cyanopropyl-25%-phenyl methyl siloxane DB-225. Split injection to a 30m x 0.25mm column coated with a 0.25 m bonded film, hydrogen carrier at 46cm/sec, isothermal at 200 °C. Components: 1, 14:1 methyl myristoleate; 2, 16:1 trans methyl palmitelaidate; 3, 16;1 cis methyl palmitoleate; 4, 18:1 trans methyl elaidate; 5, 18:1 cis methyl oleate; 6, 18;2 trans methyl linolealaidate; 7, 18:2 cis methyl linoleate; 8, 18:3 methyl linoleate; 9, 20:1 methyl cis-5-eisosenoate; 10, 22:1 methyl erucate.

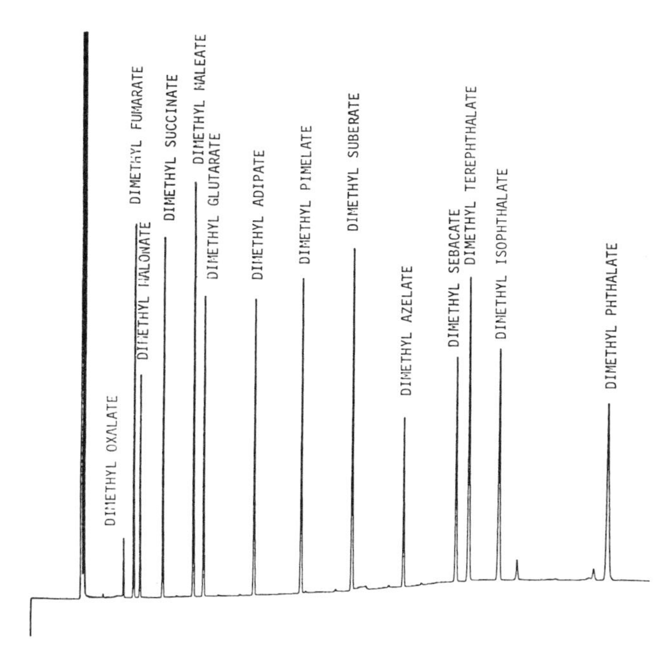

Figure 8. Dimethyl esters of the dicarboxllic acids as resolved on the dicyanoallyl silicone DB-275. Split injection to a 30 m x 0.25 mm column with a 0.15 µm bonded film of DB-275. Hydrogen carrier at 40 cm/sec; 110°C 1 min, 4°/min to 150. Components as shown.

Acknowledgement

The author is grateful to R. M. Lautamo, J & W Scientific, for the chromatograms shown in Figures 4-8.

Literature Cited

1. Jennings, W. "Gas Chromatography with Glass Capillary Columns", second edition; Academic Press: New York, San Francisco, London, 1980.
2. Desty, D. H.; Goldup, A.; Whyman; B. H. F. J. Inst. Pet., London 1976 45, 287.
3. Jennings, W.; Settlage, J. A.; Miller, R. J. HRC & CC 1979, 2, 441.
4. Jennings, W.; Settlage, J. A.; Miller, R. J.; Raabe, O. G. J. Chromatogr. 1979, 186, 189.
5. Ingraham, D. F.; Shoemaker, C. F.; Jennings, W. HRC & CC 1982, 5 227.
6. Jenkins, R.; Wohleb, R. (1980); paper presented at the 15th International Symposium "Advances in Chromatography", October 6-9, Houston, Texas.
7. Maier, H. J.; Karpathy, O. C. J. Chromatogr. 1962, 8, 308.
8. Laub, R. J.; Purnell, J. H. Anal. Chem. 1976, 48, 799.
9. Ingraham, D. F.; Shoemaker, C. F.; Jennings, W. J. Chromatogr. 1982, 239, 39.
10. Takeoka, G.; Richard, H. M.; Mehran, Mehrzad; Jennings, W. HRC & CC (1983) in press.
11. Schutjes, C. P. M.; Vermeer, E. A.; Rijks, J. A.; Cramers, C. A. J. Chromatogr. 1982, 253, 1.

RECEIVED October 14, 1983

4

Approaches to Ultrahigh Resolution Chromatography

Interactions Between Relative Peak (N), Relative Retention (α), and Absolute Retention (k')

C. H. LOCHMÜLLER

Paul M. Gross Chemical Laboratory, Duke University, Durham, NC 27706

The lack of a precise definition for "ultra-high resolution chromatography and the common confusion of resolution with plate number is discussed. Examples are given in which high resolution separations are achieved through a combination of chemical and physical insights. Cases in which plate number, relative retention and control of capacity are the dominat factors in producing high, practical resolution are discussed.

The term "ultra-high resolution chromatography" lacks a precise definition because the terms resolution and efficiency are often used interchangably. Ultra-high resolution conveys to many the concept of a separation system which is capable of producing large numbers of effective or theoretical plates. This is not suprising since high plate numbers are often easier to achieve than is high selectivity in the form of larger values of relative retention (α). The former is a problem in mechanical design whereas the latter requires more chemical insight and intuition.

This paper will illustrate, through examples drawn from our research laboratories and those of others, the interaction that exists between the three factors which determine resolution: relative peak or "band" width(proportional to N), relative retention (α) and absolute retention (k'). It is a fact, for example, that a column developing an infinite number of plates will not resolve a "critical pair" whose absolute retention is identical. On the other hand, a large value of relative retention (α) permits otherwise difficult separations to be achieved in very poor plate performance columns (Part of the secret of the success of chromatography!). The tendency to relate high-resolution to

0097-6156/84/0250-0037$06.00/0

plate number has other influences in the practice of chromatography. Most high effective plate number columns have very low sample capacity. This has two effects: 1) The amount of sample that can be applied is small and limits the use of such columns to minor component analysis. 2) Many components elute with fractional capacity factors which places an even higher plate number requirement for good resolution. In addition, the generation of large numbers of plates can increase the analysis time which decreases sample throughput. The easiest separations are between chemical classes with the order of difficulty increasing dramaticall as one progresses to within class and then to isomeric compounds (geometric<isotopic<optical). Optical enantiomers cannot be separated on any achiral column and are not separated on "just any chiral column" either.

One needs only to consult the texts in chromatography to find numerous formulations of resolution in terms of plate number, (α) and k'. While some of these formulations are more general than others and may include analysis time as a variable, the conclusion one immediately comes to is that **practical resolution** is not just a question of any one variable but an interaction between all three (or four if time is important). Ultra-high resolution requirements in an experiment may require careful planning and a significant amount of chemical intuition.

Enhancement of Plate Number

The enhancement of plate number is synonymous with minimizing relative peak or "band" width. This is first and foremost achieved through column engineering and manipulation of the physical principles of mass transfer and dispersion. Historically, column design outstrips instrument performance. For example, the use of small diameter packed columns for hplc is not new and, as Scott has pointed out, early hplc columns were actually producing 4-5 times the plates realized in the final results (1). As columns become more efficient, inlets and detectors become more important in terms of their contribution to band broadening. A good example is in the application of capillary columns in gas-liquid chromatography.

It is relatively easy to limit the detector contribution to band broadening in capillary glc in the case of the mass-sensitive, flame ionization detector or FID. Since the FID is a mass-sensitive dtector one can move the column outlet until it is practically in the flame of the FID jet. In addition, make-up gas or additional carrier or fuel flow can be introduced within the detector to rapidly sweep eluting band from the column tip to the flame. Most other detectors are concentration dependent. Since concentration detector response is well known to be inversely proportional to flow, make-up gas has a double effect one of which is undesirable - poorer detectability. Figure 1 shows the dramatic effect that systematic volume reduction and flow path redesign can have on the observed peak shape obtained with capillary columns and a thermal conductivity detector or TCD

The TCD is a concentration dependent detector like the FPD,ECD and PID but poses additional problems in that simple volume reduction is not a favorable optimization route because the response is a complex function of filament resistance, cell diameter and temperature gradient. Beginning in 1976, Lochmüller and Gordon (2,3,4) conducted a quantitative study of the effect of cell volume and flow geometry in TCD's with an eye towards a practical compromise between cell distortion of the eluting band and detectability. The result of this study was a small volume (ca. 30 ul) cell of the "Y" flow path type with four filaments which competed quite favorably with a state-of-art FID in terms of measured plates in a chromatogram at flow velocities only a factor of 2 greater than the van Deemter minimum of the FID. A comparison plot is shown in Figure 2(a) along with a temperature programmed separation of gasoline components using the cell in Figure 2(b). Similar considerations by others have led to significant improvement in detectors for hplc as well.

A particularly interesting experiment is shown in Figure 3. It is a practical observation that changing the stationary phase in a capillary column in order to enhance the separation of an unresolved critical pair can result in simply reducing the resolution of another critical pair which was originally well-resolved. On the other hand, the continued lengthening of a column to achieve better resolution through increased plates increases analysis time. Figure 3 is a pair of chromatograms taken with a recently introduced instrument which has the capability of diverting part of a developing chromatogram into an adjacent oven and column. That "cut" then is separated on a more optimum column and stationary phase. Such efficient transfer is the result of carefull design. The other important point made in this experiment is that a few single peaks in a high- resolution experiment may actually be further resolved into a significant number of individual components.

Enhanced Selectivity by Stationary Phase Modification

Three examples are presented here: 1) An example of the enhancement of resolution in the gas chromatographic resolution of enantiomers achieved by specifically-designed stationary phases or specifically selected "states" of stationary phases. 2) The use of mobile phase additives which actually

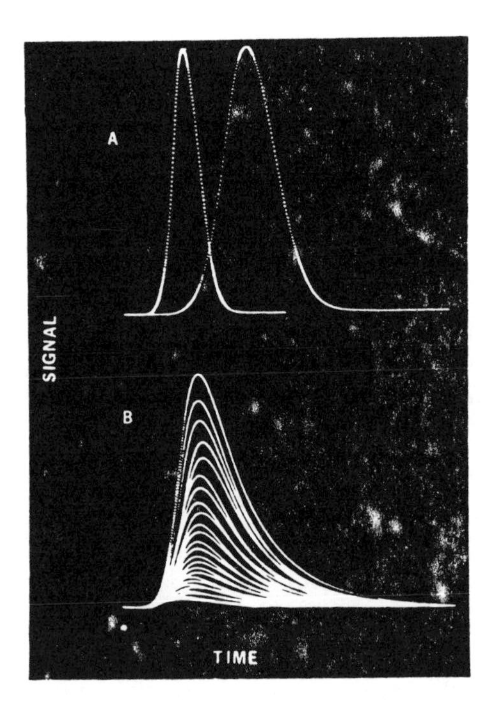

Figure 1. A comparison of the band shape obtained with a conventional "micro" TCD and a systematically designed small volume cell. Note the improvement both in width and the reduction in "tailing". (Methane peak, digitized data) Reproduced, with permission, from Ref. 1. Copyright 1977, Journal of Chromatographic Science.

a

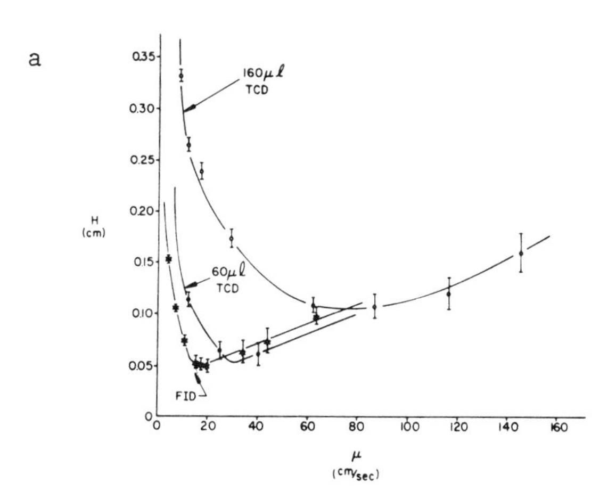

b

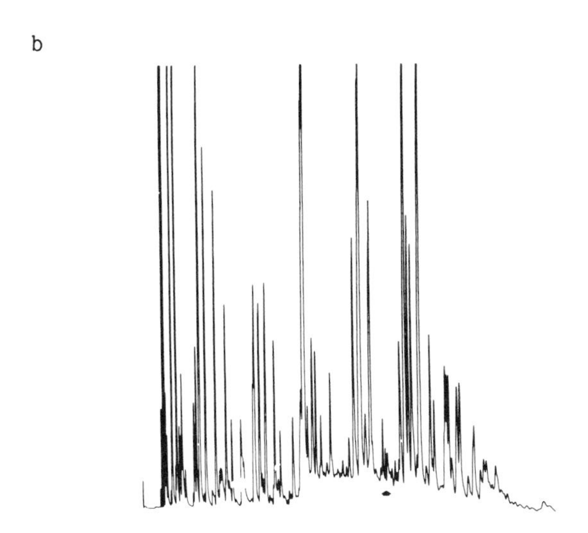

Figure 2. (a) Comparison Plate Height vs. Flow velocity for FID and optimized TCD showing equivalent performance in capillary work at practical velocities. (b) Temperature programmed separation of gasoline components using optimized TCD.Reproduced, with permission, from Ref. 2,3 Copyright 1978, J. Chromatographic Science.

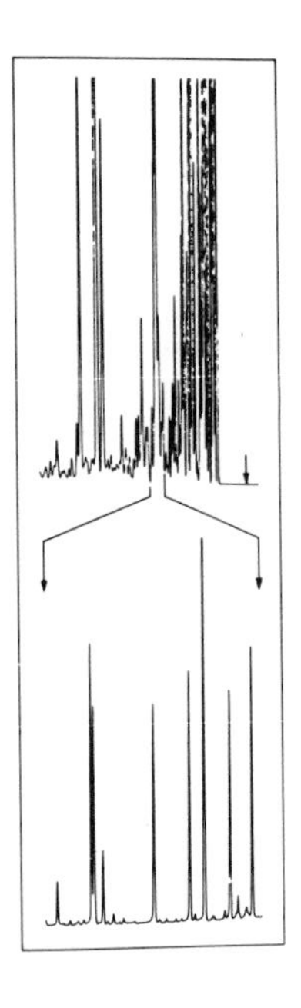

Figure 3. A separation achieved with two capillary columns of high plate number by "cutting" the eluting peaks and transferring them to a second column in a separate oven. Reproduced with permission Siemens, Copyright 1983.

provide a new, dynamic and selective stationary phase for the reversed-phase hplc separation of nitrogen bases. 3) The separation of nitro-polynuclear aromatic hydrocarbons from their parent hydrocarbons as a class on a specifically chosen bonded phase. All of the examples are for systems which ordinarily show "ultra-high resolution" in terms of plate numbers but poor practical resolution because of limited relative retention of "critical pairs." In addition, it is demonstrated that practical resolution is be enhanced by the manipulation of capacity factor of selected solutes.

Enantiomers - substances which differ only in their "handedness" and in no other physical property - are inseparable on conventional, optically-inactive stationary phases such as are commonly used for glc. Since the vapor pressure of both enantiomers is the same, separation is possible only if transient diastereomeric complexes are formed in the stationary phase. For this to happen, the stationary phase must be chiral. The use of a chiral phase alone is insufficient to guarantee the separation of enantiomers and the literature is full of failed examples. In addition to phase chirality, there is a strong requirement that the solute-solvent interactions be chiro- selective. This means that the main interactions must include ones in which the essential difference in handidness is expressed and the more this is true, the larger the difference in retention that will be observed. There has been substantial progress in this area since the late 1960's when the first authentic separations began to be reported and when relative retention values of 1.008 were common and plate requirements >105 were commonplace. Examples abound in which relative retention values of 3.0 or greater have been observed which reduce the plate requirement significantly.

The first example show the separation of enantiomeric amides on a chiral stationary phase (4(a)) and on the same phase but in its liquid-crystalline state (4(b)) (first discovered by Lochmüller and Souter (5)). The enhanced selectivity of the latter is so great that a separation which required about 35,000 plates on a capillary column is reduced to a 100 plate packed column experiment. As a result the practical capacity is increased fom the picogram to the milligram range.

The second example is a hplc experiment in the reversed-phase mode involving the separation of nitrogen bases. Here the stationary phase selectivity is altered for some bases by the addition of a nickel complex to the mobile phase. As demonstrated by Lochmüller and Hangac (6) these uncharged, coordinatively unsaturated complexes interact selectively with some bases but not others. Enhanced resolution is acheived at low concentrations (.0001 M) because some bases undergo a 10-fold increase in retention. The result is

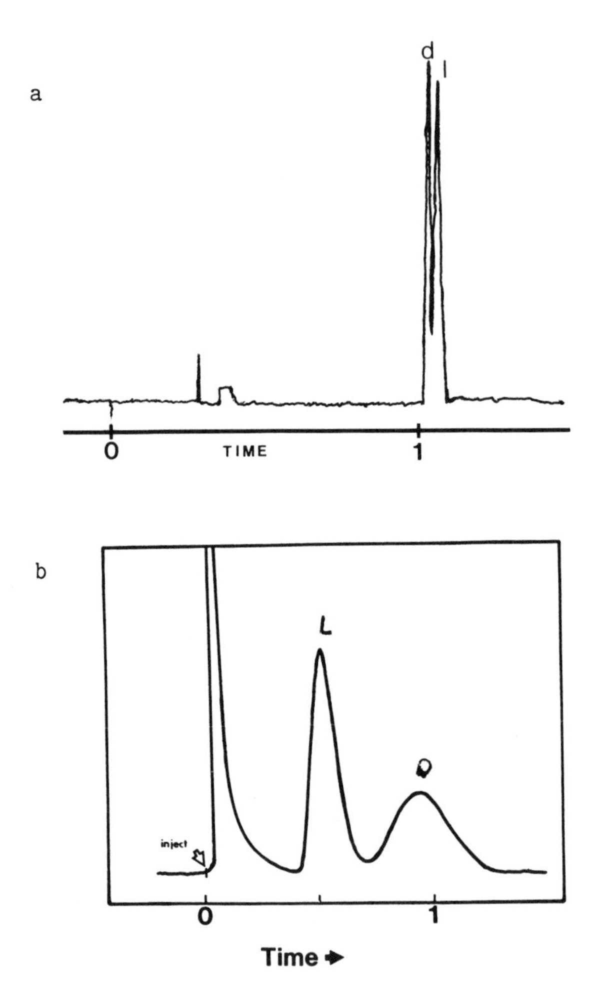

Figure 4. Separation of enantiomeric amides on carbonyl bis(D-leucine isopropyl ester) (a) in the normal liquid state of the phase and (b) when the phase is a liquid crystal. Time axis in both cases is in hours.

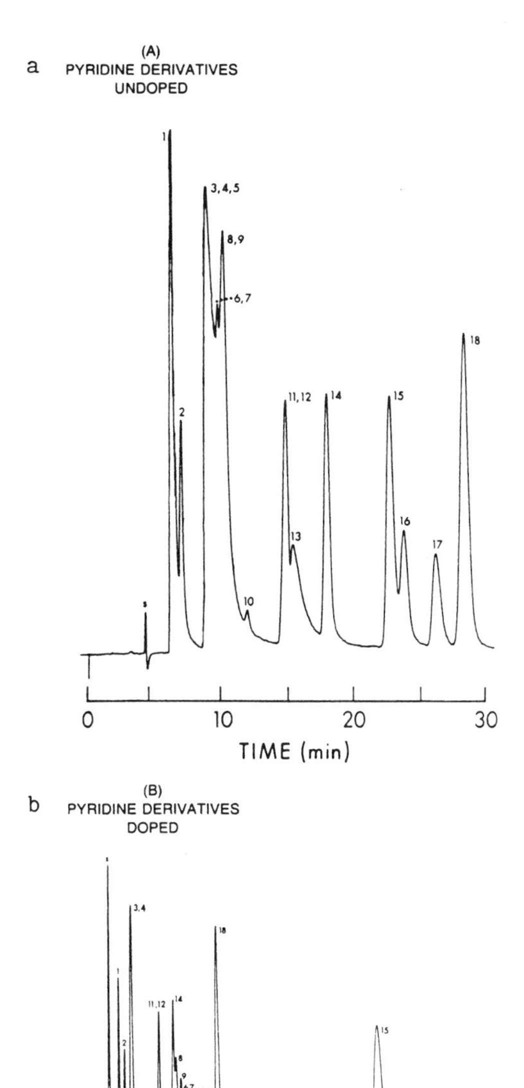

Figure 5. Separation of a variety of nitrogen bases and other aromatic heterocycles in the absence of (a) and presence of (b) nickel dipivoylmethane as a mobile phase additive in rp-hplc. The selective effect on the capacity factors is demonstrated by keeping the numbering system identical in the second chromatogram. Reproduced with permission from Ref. 6. Copyright 1982 Journal of Chromatographic Science.

a net increase in relative retention and better separation. A comparison is shown in Figure.5(a,b).

The third example is the use of a custom bonded phase in normal phase mode in hplc. The separation of the nitro-aromatics from their aromatic parent compounds is an area of significant concern because of the suspected mutagenic character of the latter which may be much greater than the aromatics themselves. Using the information gathered in a number of studies involving the separation of closely related polynuclear aromatic compounds (7,8,9) utilizing charge-transfer acceptor phases (nitro-aromatic bonded phases), Lochmüller and Beaver (10) rationalized that a bound pyrene moiety would provide high selectivity for nitro- over non-nitroaromatics. A chromatogram showing the excellent between-class and within- class separation developed by a column packed with such a phase is shown in Figure (6.) The importance of this example is the enhancement of practical resolution by increased absolute retention of the nitro species. On such a column the four-fused-ring aromatics (e.g.-chrysene) pre-elute the first nitroaromatic - nitrobenzene!

Conclusion

The achievement of "ultra-high" resolution conditions is a complex interaction between column efficiency, relative and absolute retention. In most cases, it is easier to increase plates if time and practical capacity are not important. Selectivity enhancement through choice of stationary phase can be a simplifying approach for difficult separations where interest is in certain critical pairs and where practical capacity is deemed important.

Acknowledgments

The support of this work, in part, by National Science Foundation grant CHE-8119600 is gratefully acknowledged.

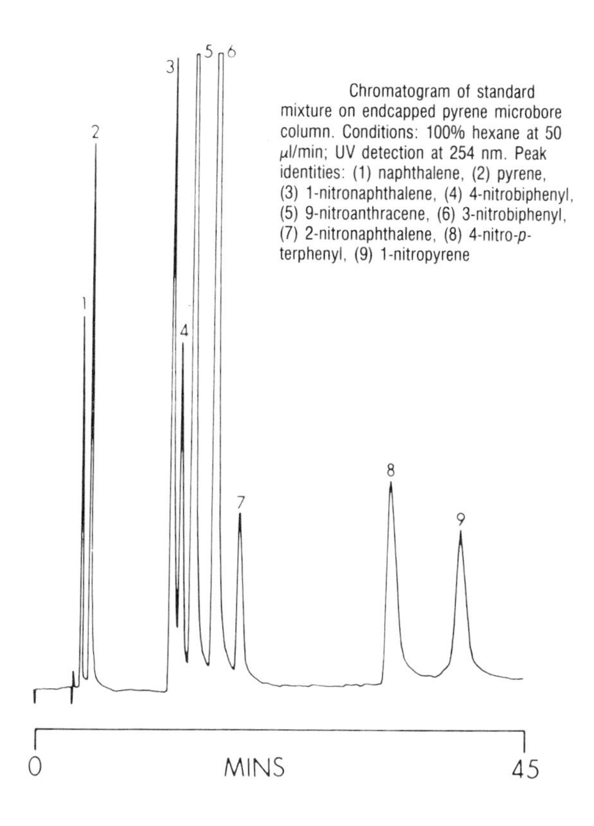

Figure 6. The separation of a test mixture of polynuclear aromatic and nitro-polynuclear aromatics using normal-phase conditions and a pyrene bonded phase. Reproduced, with permission, from Ref. 10. Copyright 1983 Journal of Chromatographic Science.

Literature Cited

1. R. P. W. Scott, 21st Eastern Analytical Symposium, New York City, November 1982.
2. C. H. Lochmüller, B. M. Gordon, A. E. Lawson and R. J. Mathieu, J. Chromatog.Sci., 15, 285 (1977).
3. C. H. Lochmüller, B. M. Gordon, A. E. Lawson and R. J. Mathieu, J. Chromatog. Sci., 16, 141 (1978).
4. C. H. Lochmüller, B. M. Gordon, A. Lawson and R. J. Mathieu., J. Chromatog. Sci., 16, 523 (1978).
5. C. H. Lochmüller and R. W. Souter, J. Chromatog., 87, 243 (1973)
6. C. H. Lochmüller and H. H. Hangac, J. Chromatog., 20, 171 (1982)
7. C. H. Lochmüller and C. W. Amoss, J. Chromatog., 108, 85 (1975)
8. C. H. Lochmüller, R. R. Ryall and C. W. Amoss, J. Chromatog., 178, 298 (1979)
9. C. H. Lochmüller and R. R. Ryall, J. Chromatog., 150, 511 (1978)
10. C. H. Lochmüller, M. Hunnicutt and R. Beaver, J. Chromatog. Sci., 21, 444 (1983)

RECEIVED December 29, 1983

5

Reduction of the Column Diameter in Capillary Gas Chromatography

Some Theoretical Aspects and Practical Consequences

B. J. LAMBERT, R. J. LAUB[1], W. L. ROBERTS, and C. A. SMITH

Department of Chemistry, San Diego State University, San Diego, CA 92182

Open-tubular columns utilized in gas chromatography have with few exceptions been of 0.2 mm diameter or larger and have therefore provided at most only a few thousand effective theoretical plates per meter length. Nevertheless, the theoretical background for design of column efficiencies of 10^5 or more effective plates per meter was presented as early as 1958. The studies reported in this work explore the practical limits of the early theoretical work and in particular, the fabrication and properties of capillary columns of inner diameter ranging from 0.3 to 0.035 mm. The latter exhibits on the order of 2×10^4 N_{eff}/m for k' of 17.

Resolution is achieved in any form of chromatography by combining selectivity with sufficient efficiency to achieve the desired separation. This amounts in principle to minimizing the kinetic factors that govern mass transport (i.e., low plate height) while maximizing the thermodynamic factors that govern partitioning (high alpha). However, superimposed on these are practical constraints such as time of analysis, detection limits, sample throughput, and so forth, all of which are embodied in the familiar resolution equation cast in terms of N, α, and k'.

Efforts at quantitative adjustment of the system selectivity, which date back to the earliest work in gas chromatography, have been reviewed most recently by Laub (**1**). Theoretical and practical aspects of column efficiency were also considered thoroughly more than twenty years ago by Golay (**2,3**), followed by Desty and his coworkers (**4-6**), Dijkstra and de Goey (**7**), Purnell and Quinn (**8**), Scott and Hazeldean (**9**), Purnell (**10**), and many others thereafter. (Reprinting of these early papers in a collected edition in fact bears consideration since virtually all are contained in volumes which are long out of print.)

However, while the practical utility of open-tubular gc columns is now widely recognized, those fabricated in laboratories as well as available commercially have until very recently generally been of 0.2

[1] Author to whom correspondence should be addressed.

0097-6156/84/0250-0049$06.00/0

mm or larger in diameter and so, have provided at most about 4000 effective theoretical plates per meter N_{eff}/m. Broadly speaking, the choice of this particular diameter can in large measure be attributed to a compromise between column efficiency, limits of detection, ease of fabrication and, in the case of fused amorphous silica systems, cost and availability of tubes of appropriate uniformity and length. The recent introduction of highly sensitive detectors (including compact and inexpensive mass spectrometers) has largely obviated difficulties associated with sample size (i.e., trace analysis), while empty capillary tubing of virtually any dimension can now be obtained commercially. Nevertheless, it is not entirely clear even today with what functions efficiency and time of analysis can be related to internal diameter at constant stationary-phase film thickness or volume, nor what demands are placed in practicable terms upon the gc instrument when some "ultimate" level of user-defined efficiency is sought. Moreover, there remains the question of ease of fabrication of systems of very small diameter and in addition, to what extent the efficiency of such columns approaches limits predicted by theory. We sought in this work to explore selected aspects of these areas, and recount here our experiences in the matters of column fabrication and evaluation.

THEORY

The Golay equation for open-tubular columns of circular geometry is given by:

$$H = \frac{2D^{o}_{1M}}{u_o} + \left[\frac{1 + 6k' + 11(k')^2}{24(1 + k')^2}\right]\left(\frac{r^2}{D^{o}_{1M}}\right)u_o + \frac{2}{3}\left[\frac{k'}{(1 + k')^2}\right]\left[\frac{(d_f)^2}{D_{1S}}\right]\bar{u} \tag{1}$$

where D^{o}_{1M} and D_{1S} are the solute diffusion coefficients in the mobile and stationary phases, respectively (the former being that referenced to the column outlet pressure), d_f is the stationary-phase film thickness, r is the tube radius, and u_o and $\bar{u}$ are the outlet and average linear mobile-phase velocities. It can be shown that minima will occur in plots of H against u_o at 0.58r and at 1.91r for solutes of $k' = 0$ and $k' \gg 0$, respectively, i.e., in the limits of negligible to large mass-transfer non-equilibrium. Thus, we have to hand means of assessing system efficiency in terms of column radius, the solute capacity factor, and the resultant experimental plate height. Further, since the capacity factor is related directly to the partition coefficient and stationary- and mobile-phase volumes, viz., $k' = K_R V_S/V_M$, we can in addition predict the effects that alteration of the ratio of phases will have upon H_{min}.

EXPERIMENTAL

Materials and Reagents

The four sizes of capillary tube (all of glass type KG-33) utilized in this work were from Ace Glass. The ranges of dimensions of each were found to be:

OD/mm	Bore/mm
5-7	0.25-0.75
6-7	0.75-1.25
6-7	1.00-1.50
6.5-7.5	2.50-3.00

Hexamethyldisilazane (HMDS) was obtained from Fluka, SE-30 polydimethylsiloxane stationary phase was purchased from Applied Science, and decane solute was from Chem Service. All were used without further purification.

Apparatus and Equipment

A Shimadzu GDM-1 glass-drawing machine was modified in our laboratory in approximate accord with the work of Schenning, van der Ven, and Venema **(11)**.

A Carlo Erba Model 4160 gas chromatograph equipped with split/splitless injection systems and a flame ionization detector was employed throughout this work. Chromatograms were recorded with a Linear Model 55 strip-chart recorder. Hydrogen was the carrier gas in all instances.

Column Preparation

Column preparation was carried out with an SGE coating reservoir and a water-bath consisting of a 15 x 30 cm diameter Dewar flask for thermostating columns during the static coating process, where a single-stage mechanical vacuum pump was used to remove coating-solution solvent from the columns. It was found that the coating reservoir fittings would not sustain the pressure required to force coating solution into columns of 0.1 mm ID and smaller and so, a high-pressure reservoir was fabricated from a stainless-steel cylinder containing a glass insert and fitted with the septum holder from a Pye Model 104 gas chromatograph. The unit was used at pressures of in excess of 700 psig without failure. The volume and average internal diameter of drawn capillaries were determined by filling and weighing with carbon tetrachloride. The remaining column-fabrication procedures were as described by us previously **(12)**.

RESULTS AND DISCUSSION

Presented in Table I are the dimensions and properties of the several columns prepared in this work. The internal diameter was varied as shown from 0.278 mm to 0.0345 mm, while the stationary-phase film thickness was held constant for all but the last of the columns. Thus, the phase ratio (V_M/V_S) decreases on passing from column 1 to 6 over the range 231.5 to 54.25. The solute capacity factors increase accordingly from 6.83 to 29.47. The number of theoretical plates per meter length N/m for all columns except the first is therefore very nearly equal to the number of effective theoretical plates per meter, since the capacity factors are close to or exceed 10.

Table I. Dimensions and Properties of Capillary Columns of This Work[a]

Col. No.	ID/mm	d_f/um	V_M/V_S	k'	$\bar{u}_{opt}$/ cm sec^{-1}	N/m
1	0.278	0.30	231.5	6.83	41.6	3770
2	0.238	0.30	197.5	8.50	40.5	4190
3	0.191	0.30	158.1	10.18	44.6	4850
4	0.149	0.30	123.0	13.12	51.5	6430
5	0.107	0.30	88.42	18.48	43.1	8590
6	0.066	0.30	54.25	29.47	35.7	12,750
7	0.0345	0.10	85.50	17.24	31.4	21,000

[a]decane solute at 65°C; hydrogen carrier; SE-30 stationary phase

Efficiency

The number of theoretical plates generated by the columns investigated here was indeed a linear function of the inverse of the column diameter, as shown in Figure 1 for columns 1-6. There was some fall-off in efficiency with the seventh column, which was more than likely due to non-uniformity of the capillary tube as well as uneven film deposition of the stationary phase. Considerable difficulties were in fact encountered initially in achieving sufficient temperature control of the water-bath in order to prevent the coating solution from expanding and contracting. A Tamson circulating water-bath of double-wall construction was finally resorted to, the stability of which was estimated to be better than $\pm$ 0.001°C. The detector sensitivity was also found to be inadequate even when increased by a factor of two beyond its maximum range with an external amplifier. At this sensitivity, the solute peak still fronted slightly, but the noise level was approaching unacceptable levels and no further reduction in sample size (split ratio of 10,000:1) was possible. Nevertheless, comparison with H_{min} predicted by the Golay equation is encouraging: 10^5 plates per meter are expected for a solute of negligible capacity factor with a column of dimensions comparable to column 7, while 30,500 N/m should be found for a solute of very large k'. Our result of 21,000 N/m thus falls within 70% of that predicted by the theory, a satisfying achievement given the fact that only commercially-available chromatographic equipment and a strip-chart recorder were utilized. Even so, it is not difficult to imagine that given the imminent availability of more sensitive gc-dedicated detectors (e.g., compact mass spectrometers), much higher efficiencies will soon be realized on a routine basis. In the meantime, and on the basis purely of practicality, ca. 0.1 mm (10,000 N_{eff}/m for k' = 10) most likely represents the region of optimal column internal diameter.

Capacity Factor and Analysis Time

The solute capacity factors measured at the optimum linear carrier velocity increase monotonously as the inverse of the column diameter, as shown in Figure 2. This result is to be expected, since the phase ratio decreases in the same fashion. However, since the mobile-phase volume decreases geometrically upon reducing the column radius, raw analysis times for a given linear carrier velocity, while longer with the columns of smaller ID, are not inordinately so. The values of t_R/m (min m^{-1}) for columns 1 to 6 respectively were 2.38, 2.56, 2.62, 2.67, 3.03, 3.63, while column 7 gave 3.14 (see below).

Linear Velocity

We see from Equation 1 that mass-transfer non-equilibrium arising from solute interaction with the stationary phase depends upon r^2/D^o_{1M}. As a consequence, there is a maximum to be expected in plots of $\bar{u}_{opt}$ against column internal diameter since, as shown above, k' increases linearly as the radius is reduced. This effect was first postulated by Purnell [Figure 6 of **(13)**] in a different context (droplet formation with packed columns, which is entirely analogous to film thickness with capillary systems) but to our knowledge has yet to be verified experimentally. The relevant data from the present study are shown plotted in Figure 3, which in fact exhibit a distinct maximum in the optimum linear carrier velocity at a column of diameter of ca. 0.15 mm ID. Moreover, this portends that for a fixed film thickness there exists for a given separation an **optimal** column radius which will require the highest linear carrier velocity and, hence, yield the fastest overall analysis time. Nor is the gain likely to be trivial since we see from the plot that $\bar{u}_{opt}$ for a column at the curve maximum may well be a factor of two higher than that either for column 1 or for column 6. In fact, the analysis time of the test solute was very nearly identical with columns 3 and 4 even though there was a 20% difference in the respective internal diameters.

Inlet Pressure

The price to be paid for utilizing columns of small ID is illustrated graphically in Figure 4, which shows the column inlet pressure per meter

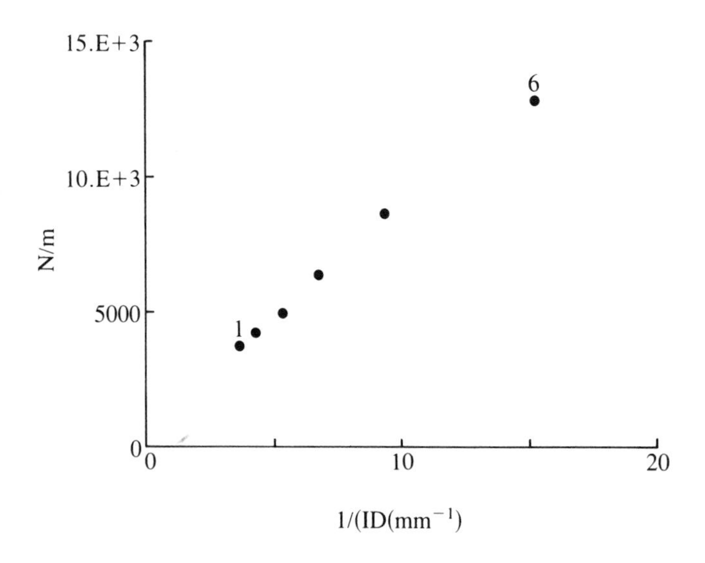

Figure 1. Plot of theoretical plates obtained per meter length against reciprocal of internal diameter at $\bar{u}_{opt}$ for columns of Table I.

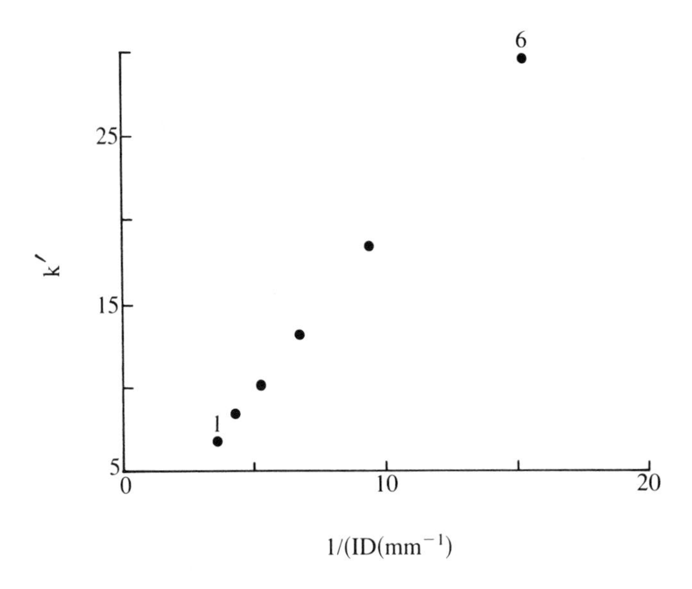

Figure 2. Plot of test-solute (n-decane) capacity factor against reciprocal of internal diameter at $\bar{u}_{opt}$ for columns of Table I.

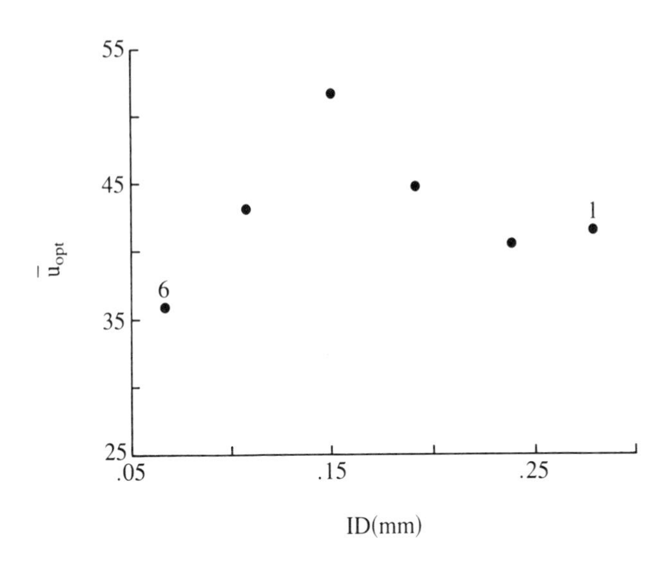

Figure 3. Plot of $\bar{u}_{opt}$ against internal diameter for columns of Table I.

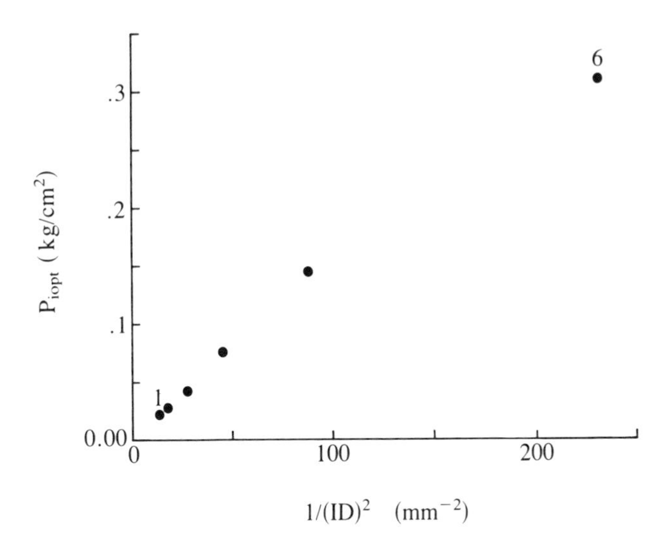

Figure 4. Plot of inlet pressure per meter length at $\bar{u}_{opt}$ against inverse square of internal diameter for columns of Table I.

length required to achieve the optimum linear carrier velocity plotted against the inverse square of internal diameter. (Parenthetically, we found that reducing the data to this form of linear regression was advantageous in practice in approximating the requisite p_i for new columns.) Furthermore, most commercial gas chromatographs are designed to operate at pressures of up to only 5 kg cm^{-2}. With some modification, this could easily be raised to ca. 15 kg cm^{-2} and in fact to do so (as was required for column 7) proves in most instances to be simply a matter of bypassing or replacing the existing instrument pressure regulator and gauge. Moreover, virial effects are trivial with hydrogen and helium carriers even at pressures of well in excess of 50 kg cm^{-2} (although the same may not be true for carrier solubility in the stationary phase).

Example Chromatograms

Figures 5(a) and (b) illustrate the chromatograms obtained at $\bar{u}_{opt}$ with columns 1 and 7, respectively, of a commercial petroleum product ("white gas"). The temperature was programmed but the rate was virtually identical in each case. There are approximately a third again as many peaks visible in the latter than in the former and, furthermore, all are obviously considerably sharper (smaller baseline widths). The last-eluting compound provides an example: the peak is barely discernible in Figure 5(a) at an elution time of 30 min, but is clearly seen at 36 min in 5(b). In addition, several very small peaks are visible penultimate to this, but are not detected in 5(a). Overall, then, the column of reduced internal diameter and film thickness is very much superior as expected on the basis of efficiency. Nevertheless, whether the results obtained justify the practical difficulties associated with its fabrication and use remains a question open largely to subjective judgment.

ACKNOWLEDGMENTS

Support of this work provided in part by the Department of Energy and by the National Science Foundation is gratefully aknowledged.

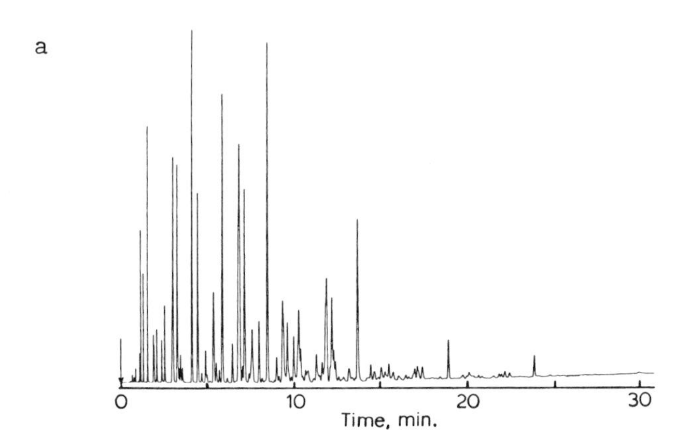

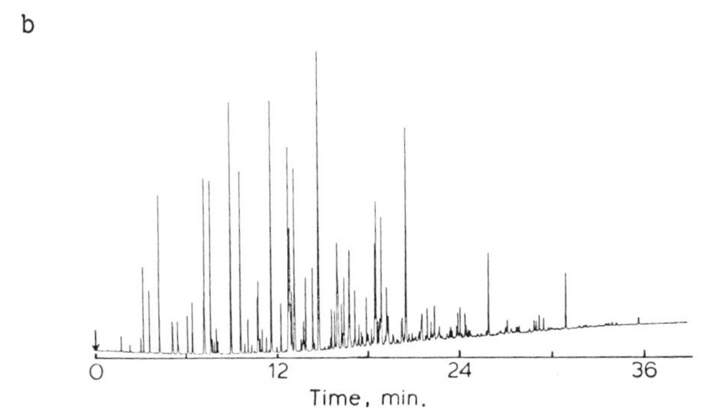

Figure 5. Chromatograms of commercial petroleum product with columns 1 (a) and 7 (b) of Table I. Temperature held in each case at 0° for 2 min, and then programmed from 0° to $70^{o}C$ (a) or to $85^{o}C$ (b) at 3° min^{-1}.

LITERATURE CITED

1. Laub, R. J. in **Physical Methods in Modern Chemical Analysis**; Kuwana, T., Ed.; Academic Press: New York, 1983; Chap. 4.
2. Golay, M. J. E. in **Gas Chromatography**; Coates, V. J., Noebels, H. J., and Fagerson, I. S., Eds.; Academic Press: New York, 1958; 1.
3. Golay, M. J. E. in **Gas Chromatography 1958**; Desty, D. H., Ed.; Butterworths: London, 1958; 36-53.
4. Desty, D. H.; Goldup, A.; Whyman, B. H. F. **J. Inst. Petrol.** 1959, **45**, 287-298.
5. Desty, D. H.; Goldup, A. in **Gas Chromatography 1960**; Scott, R. P. W., Ed.; Butterworths: London, 1960; 162-181.
6. Desty, D. H.; Goldup, A.; Swanton, W. T. in **Gas Chromatography 1962**; van Swaay, M., Ed.; Butterworths: London, 1962; 105-135.
7. Dijkstra, G.; de Goey, J. in **Gas Chromatography 1958**; Desty, D. H., Ed.; Butterworths: London, 1958; 56-62.
8. Purnell, J. H.; Quinn, C. P. in **Gas Chromatography 1960**; Scott, R. P. W., Ed.; Butterworths: London, 1960; 184-198.
9. Scott, R. P. W.; Hazeldean, G. S. F. in **Gas Chromatography 1960**; Scott, R. P. W., Ed.; Butterworths: London, 1960; 144-159.
10. Purnell, J. H. **J. Chem. Soc.** 1960, 1268-1274.
11. Schenning, J. G.; Van der Ven, L. G. J.; Venema, A. **J. High Resolut. Chromatogr. Chromatogr. Commun.** 1978, **1**, 101-102.
12. Laub, R. J., Roberts, W. L., and Smith, C. A. **J. High Resolut. Chromatogr. Chromatogr. Commun.** 1980, **3**, 355.
13. Purnell, J. H. **Gas Chromatography**; Wiley: London, 1962; p. 148.

RECEIVED September 12, 1983

6

Use of a Small Volume Thermal Conductivity Detector for Capillary Gas Chromatography

Practical Aspects

RICHARD SIMON and GREGORY WELLS

Walnut Creek Division, Varian Instrument Group, Walnut Creek, CA 94598

Capillary chromatography is attractive for analysis of complex samples, primarily because the capillary columns have very high efficiencies when compared with packed columns. In practice however, capillary columns are not operated at maximum efficiency, and a trade-off is made to reduce analysis time by sacrificing chromatographic resolution.

The use of capillary columns is a relatively recent advance in chromatography when compared with the thermal conductivity detector (TCD). The TCD is well established and is among the most commonly used detectors in gas chromatography. Some of the advantages of the TCD include the simplicity, stability, and universal nature of the detector.

This paper addresses the operating characteristics and performance of a 30 μℓ TCD when coupled with a high resolution capillary gas chromatograph. Differences in the peak shape, peak symmetry, electronic bandwidth, column flow rates, and column bleed invoke different responses for the TCD with a capillary column as opposed to a packed column. These differences raise several questions which will be addressed in this paper: (1) How does the detector volume of the TCD affect the observed chromatographic performance? (2) How is the best performance obtained in practice? (3) How do detection limits, linear dynamic range, and sensitivity change using a capillary chromatograph as opposed to a packed chromatograph?

0097-6156/84/0250-0059$06.00/0

In depth discussions of the TCD can be found in most texts on gas chromatography [1-4]. Lawson and Miller [5] have written a review of the TCD, and give multiple references to discussions regarding the TCD. A comparison of the various modes of operation has been made by Wells and Simon [6] which compares the constant voltage, constant current, and constant mean temperature modes of operation on both an empirical and a theoretical basis.

The TCD measures changes in the thermal conductivity of the carrier gas, perturbed by the eluting analyte. Thermal conductivity detectors which rely on diffusion by the analyte to the hot filament generally have the lowest limit of detection (due to the very low noise of the detector), yet have very long time constants. Flow-through thermal conductivity detectors can retain peak symmetry without excessively large volumetric time constants.

The sensitivity of the TCD is defined as the signal output per unit concentration [7], or as:

$$S = \frac{A\ F_c}{W} \qquad (1)$$

where S is the TCD sensitivity (mVmL/mg),
A is the integrated peak area (mV min),
F_c is the gas flow rate at the detector (mL/min), and
W is the weight of the sample (mg).

Sensitivity of a TCD is dependent upon the detector block temperature, the detector filament temperature, carrier gas, sample, detector volume, and detector configuration. These should be listed when specifying a sensitivity for a TCD, when comparing different instruments, or different instrumental conditions.

Since the TCD is a flow sensitive (concentration) detector, the low flow rates generally associated with capillary systems should enhance the observed response of a low volume (30 μℓ) TCD. The observed response, i.e. peak height, or peak area, is inversely proportional to the gas flow rate at the detector (assuming the same carrier gas flow rate through the column, with any additional gas being introduced as a make-up flow just prior to the detector). Optimum detector response should be obtained at those flow rates just sufficient to efficiently sweep-out the dead volumes associated with the detector and connecting tubing.

Since sharp, narrow peaks are generated with a capillary system, the detector must be efficiently purged by the carrier gas. Otherwise, the detector and connecting dead volumes operate as a mixing or an exponential dilution chamber and thereby severly distort the actual chromatographic peak shape. Electronically this is analogous to increasing the electronic time constant. This volumetric time constant (τ) is given by:

$$\tau = V_{eff}/F_c \tag{2}$$

where V_{eff} is the effective volume of the TCD, and F_c is the gas flow rate at the detector. Sternberg [8] has shown that electronic and volumetric time constants can seriously distort chromatographic peaks when τ (or RC) is greater than one fifth the standard deviation σ of the chromatographic peak.

The effect of a volumetric or electronic time constant can be modeled as a convolution of a Gaussian peak g(t) with a filter function (or transfer function) h(T). This convolution generates the output signal f(t) defined by:

$$f(y) = g(t) \ * \ h(t) \tag{3}$$

For an electronic or turbulent volumetric time constant τ, h(t) is given by:

$$h(T) = (1/\tau) \exp(-T/\tau) \tag{4}$$

For plug flow the filter function corresponds to a simple averaging of the input function and has the form:

$$h(T) = 1/\tau \tag{5}$$

Thus the output signal is of the form:

$$f(t) = \int_a^b h(T) \ \ g(t-T) \ dT \tag{6}$$

where a and b are the integration limits (for turbulent flow a = 0, b = t, and for plug flow a = t - τ , b = t). Both the exponential filter and plug filter can be rigorously solved by the method of Laplace transforms. However, the analytical function is best characterized by the use of statistical moments.

An alternative definition of peak distortion is peak fidelity or the ratio of the output peak height to the input peak height. Peak fidelity (f) can be approximated by:

$$f = \sigma^2/(\sigma^2 + \tau^2) \tag{7}$$

For a peak fidelity of 98%, a σ/τ ratio of 4.92 or greater is required. In other words, the standard deviation of the peak must be greater than 4.92 V_{eff}/F_c. For sharp, narrow peaks, the detector time constant can be reduced by reducing the effective detector volume, increasing the make-up gas flow rate, or both. However, increasing the make-up gas flow rate decreases the detector responses, and results in a loss in detectivity if the noise remains constant.

The theoretical sensitivities which can be attained with concentration detectors for both packed and capillary systems have been reported by Yang and Cram [9]. The concentration detectors will have better detectivities when using a capillary column compared with a packed column, defined by :

$$D = \frac{V_p}{V_c}\left[\frac{N_c}{N_p}\right]^{\frac{1}{2}} \tag{8}$$

where D is the detectivity ratio of capillary to packed columns, V_p is the retention volume of the packed column, V_c is the retention volume of the capillary column, and N_p, and N_c are the number of the theoretical plates for packed and capillary columns respectfully.

This expression implicity assumes the same noise sources, mixing efficiencies within the detectors, and comparable electronic time constants. This expression can be rewritten as:

$$D = \frac{\sigma_p \; F_p}{\sigma_c \; F_c} \tag{9}$$

Since $\sigma_c \ll \sigma_p$ and $F_c < F_p$, superior performance is expected for capillary systems as opposed to packed columns with a TCD.

Equation (9) shows that when a TCD is used to detect the eluent from a capillary column, it will have a lower detection limit than that for a packed column, or a higher signal to noise ratio for the same sample size injected.

Because the minimum detectable quantity for a TCD is proportional to the concentration of sample molecules at the peak maximum, C_{max}, the instantaneous concentration, C, for a Gaussian distribution in volume units is:

$$C = C_{max} \; e^{-z^2/2\sigma_v^2} \tag{10}$$

The total mass, M, can be written as:

$$M = \int C_{max} \; e^{-z^2/\sigma_v^2} \; dz \tag{11}$$

or

$$M = C_{max} \sqrt{2\pi}\ \sigma_v$$

Here σv corresponds to the gas volume associated with the peak standard deviation or in the time domain.:

$$\sigma_v = \sigma * F_c \tag{12}$$

The detection limit, or detectivity d, is often defined as :

$$d = 2N/S \tag{13}$$

where N is the noise (mV) and S is the detector sensitivity as defined in equation 1. The detectivity corresponds to the concentration of sample molecules at the peak maximum in the detector volume. Thus the total mass M which must be accomodated by the column at the detection limit is given by:

$$M = \frac{2N\sqrt{2\pi}\ \sigma_t\ F_c}{S} \tag{14}$$

When a split injection technique is used, only a fraction of the sample is actually injected onto the column. This fraction corresponds to the split ratio of the injector, χ . The split ratio is defined as the ratio of column flow rate to the split point flow rate.

Sample capacity is a measure of the maximum sample mass which can be accomodated without overloading the column. Overloading occurs when the mobile to stationary phase distribution isotherm becomes non-linear, resulting in distorted peak shapes and loss of chromatographic resolution.

The phase ratio, i.e. the ration of the volumes occupied by the gas phase and the stationary phase, is given the symbol β . For wall coated capillary columns, β is approximated by:

$$\beta = {}^{r}/2d_f \tag{15}$$

where r_o is the distance from the center of the column to the surface of the liquid phase, and d_f corresponds to the film (stationary phase) thickness. The range for β values typically vary from 5 to 35 for packed columns to 50 to 1500 for capillary columns. The effect of increasing β for a fixed column diameter and column length is to decrease the sample retention time and sample capacity of the column. Increasing the stationary film thickness results in increased sample capacity, but at the expense of longer retention times or higher temperatures. In a similar manner, increasing the column diameter, but maintaining the same stationary phase thickness, results in lower retention times but an increase in column sample capacity due to the increased stationary phase present on the column.

A conventional capillary column, 0.200mm x 12.5M with a β of 250 has a sample capacity of about 50 ng at $k' = 10$. In comparison, a 0.480 mm x 28M column with a β of 250 has a sample capacity of about 540 ng at a $k' = 10$. These very low sample capacities affect the TCD requirements very severely since the minimum detectable quantity, i.e. the detectivity, approaches the maximum sample capacity of the column as the column diameter decreases.

Experimental

The small volume thermal conductivity detector used in this study was a Y-type [10,11,12] gas flow pattern 30 µℓ Gow-Mac Model 10-955 cell (Gow-Mac Instrument Company, Madison, New Jersey). This cell was compared with the standard Varian TCD (Gow-Mac Model 10-952) in a Varian series 3700 gas chromatograph. The standard electronics were modified for operation with the 30 µℓ TCD. All measurements were made in the constant mean temperature mode. The carrier gas was He. The flow rates were regulated by two 0-60 ml/min mass flow controllers (model 1000, Porter Instrument Company, Hatfield, Pa.) or by a pressure regulator (Model 8601, Brooks Instrument Division, Emerson Electric Company, Hatfield, Pa.). The capillary column was attached to a modified 1/16" to 1/16" Swagelok union which in turn was connected to the appropriate TCD. All make-up flows were regulated so that the total flow through both the reference and the sample sides were matched.

A 0.48 mm x 28M OV-101 column (SG&E, Austin Texas) was used with a Varian 1080 injector. A 0.20 mm x 12.5M SE-54 column (J & W Scientific Inc.) was used with a Varian 1070 injector.

Distilled in glass 2,2,4-trimethylpentane (Burdick & Johnson Laboratories Inc., Muskegon, MI.) was used as solvent without any additional purification. Solutes (n-pentadecane, n-hexadecane, and n-tetradecane) were obtained from Poly Science Corporation (Niles, Il.) and were used without additional purification. All calibration curves were made by the method of sequential dilution.

Data acquisition was carried out by either a Varian 401 data system, or by a HP-1000 computer system. Peak areas, retention times, peak widths and peak heights were obtained using both systems. All data were obtained with a 20 Hz sampling rate.

Fast peaks were generated using a modified fast sampling valve (Linear Dynamics, E. Pepperell, Ma.), to inject butane (Matheson, instrument grade) onto an uncoated 0.060mm x 2.0m fused silica column. The 1070 injector cap was drilled out to allow a piece of 1/16" steel tubing to seal the fast sampling valve with the 1070 injector body. The standard gas inlet line to the 1070 was capped with a Swagelok plug and the gas line was connected to the fast sampling valve (see Figure 1). The gas flow rate was held at 300 ml/min. The split point for the injector was raised to within 1 cm of the fast sampling valve's outlet port and was constantly swept by 300 ml/min of helium. the valve was opened for 50 ms for each data run. The data was collected on a Nicolet 1170 signal averager, 10 ms per point, 1024 points per run, synchronized with the fast sampling valve. Different peak widths were generated by changing the column temperature (e.g. 90 C for 0.33 second peak width, 230 C for 1.07 second peak width, and 340 C for 1.65 second peak width).

Results And Discussion

The standard thermal conductivity detector employed by Varian is a Y-geometry 140 μℓ cell consisting of four matched filaments arranged in a Wheatstone bridge configuration and operated in the constant mean temperture mode. The cell was replaced by a 30ul TCD and the electronics were modified to compensate for change in filament resistance.

The filament resistance of the 30 μℓ TCD was determined as a function of temperature by heating and cooling the 30 μℓ TCD in an oven (regulated 0.01°C), allowing time for the TCD to thermally equilibrate prior to measuring the filament resistances. Plotting the average bridge resistance versus the oven temperature permitted fitting the 30 μℓ TCD filament resistance as a function of temperature to the form given by:

$$R(t) = R_o \ (1 + \propto T)$$

where T is the filament temperature, $\propto$ is the resistivity coefficient, Ro is the zero-point resistance, and R(T) is the resistance at filament temperature T. For the 140 μℓ TCD, Ro = 29.6 ohms, $\propto = 0.00322/^{o}C$. For the 30 μℓ TCD, Ro = 13.1 ohms, $\propto = 0.00788/^{o}C$.

Using helium as carrier gas with two matched mass flow controllers, (30 ml/min He through both sides), and operating in the constant mean temperature mode (no columns, empty stainless steel tubing only), current versus ΔT (where ΔT is the difference between the filament and wall temperatures) was measured for the 30 μℓ TCD (see Figure 2). Use of a 1/8" x 20" 5% OV-101 packed column (Varian, Walnut Creek, CA) allowed determination of sensitivity versus current as shown in Figure 3. These curves reveal that a comparison of the 30 μℓ TCD and the 140 μℓ TCD can not be made on the basis of current, but rather must be based on equal ΔT. This may seem to imply that the 140 μℓ TCD may be preferred due to greater sensitivity, but this simplification totally neglects the effects of detector volumetric time constants on peak shapes.

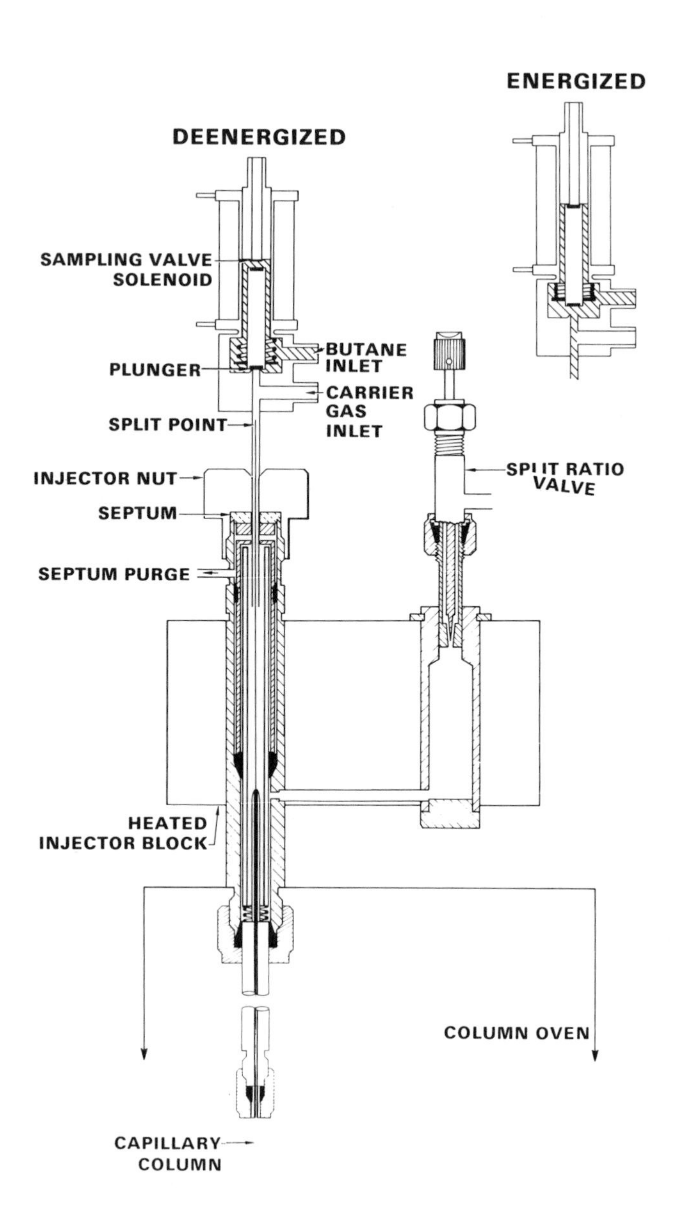

Figure 1. Fast sampling valve inlet system: Varian 1070 injector and modified Linear dynamics valve.

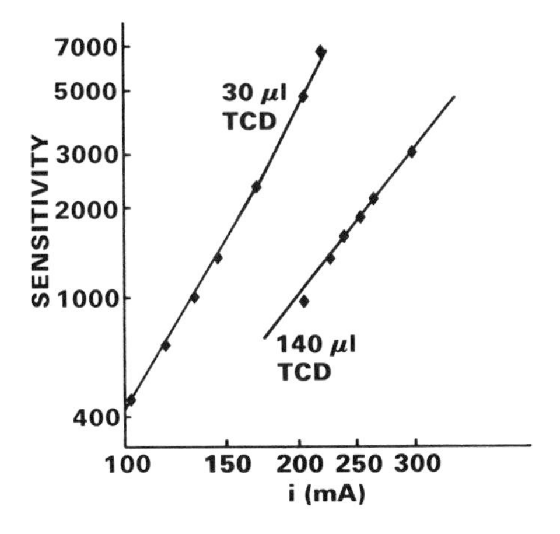

Figure 2. Current versus ΔT for 30 μℓ and 140 μℓ thermal conductivity detectors. Reference and sample flow rates 30 ml/min of helium. Detector temperature 170°C.

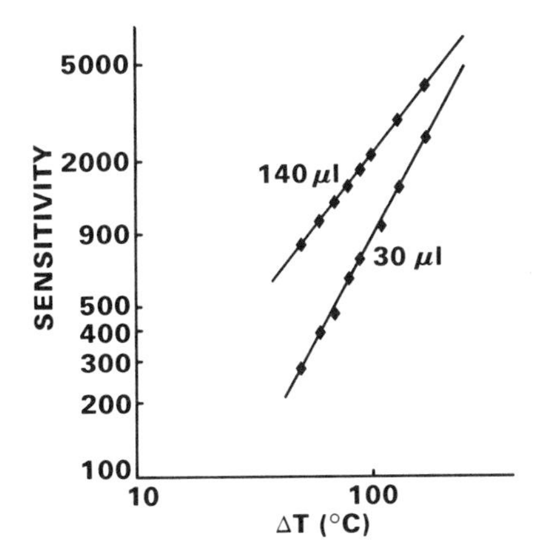

Figure 3. Sensitivity versus Δ T for 30 μℓ and 140μℓ thermal conductivity detectors. Reference and sample flow rates 30 ml/min. helium. Detector temperature 140°C. Injector temperature 180°C.

Thermal conductivity detectors generally have been used with packed columns. Connections between the column and the detector are made by using GLT (stainless steel glass lined tubing). The GLT serves to reduce void volumes at the detector-column interface, provides a completely fixed geometry which is reproducible, and self-aligning whenever the column is changed, and minimizes the solute residence time and peak mixing effects in connecting tubing. The GLT was used with capillary columns because the inner diameter of the GLT is 0.30mm (comparable with the wide bore 0.48mm glass column, and compatible with the 0.20mm fused silica column), and is swept by not only the column flow, but also any make-up flow which is being used as a purge flow, thus minimizing solute residence time, band broadening, and tailing due to connecting tubing.

As stated earlier, the TCD is a concentration dependent detector and therefore has a detector volume that must be efficiently swept by either carrier gas, or make-up gas, to prevent any significant peak distortion. Comparison of a 140 μℓ TCD and a 30 μℓ TCD, requiring equal volumetric time constants and equivalent fidelity of a Gaussian input peak shape, requires the detector flow rate for the 140 μℓ TCD to be 4.6 times larger than that for the 30 μℓ TCD. This difference in total detector flow rates allows the 30 μℓ TCD to give a greater response than the 140μℓ TCD by simply decreasing the flow rate thru the 30 μℓ TCD.

Examination of the response of the 30 μℓ TCD as a function of flow rate (see Figure 4) reveals apparent linear response for several different flow rates, and a decrease in the minimum detectable quantity proportional with the decrease in total flow through the detector. This strongly supports the concept that the TCD operates with greatest sensitivity and minimum distortion at flow greater than 4.92 V_{eff}/σ (e.g. for 30 μℓ TCD and σ =1 second, the total flow thru the detector must be greater than 8.9 ml/min, and for the 140 μℓ TCD the flow must be greater than 41.3 ml/min for minimum distortion).

Sternberg [8] has shown that the observed peak variance is the sum of the individual variances of all processes that contribute to band broadening, or:

$$\sigma^2_{total} = \sigma^2_{inj.} + \sigma^2_{col.} + \sigma^2_{det.} + \sigma^2_{ct} \qquad (16)$$

where σ^2 are the variance associated with the injector, the column, the detector, and the connecting tubing. In this method, variances are calculated in the time domain (as opposed to the column-length domain) and correspond to the second moment of the chromatographic peak, i.e. $\sigma^2 = M_2/M_o$ [8]. For a flame ionization detector (FID), σ^2 det is approximately zero, the

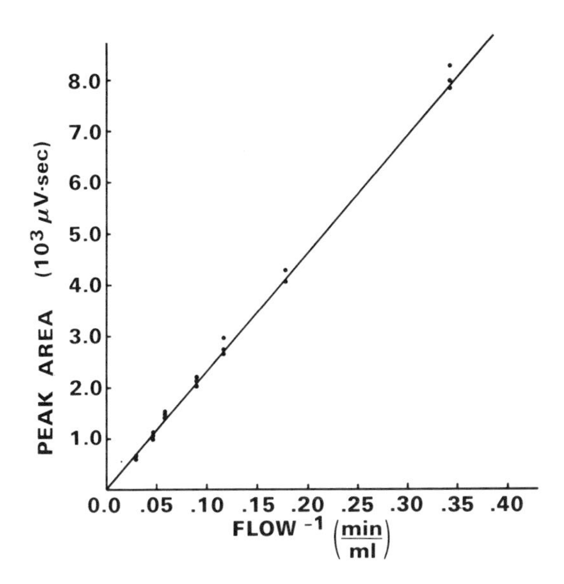

Figure 4. Response of the 30 μℓ TCD versus flow rate. Detector temperature 170°C, injector temperature 180°C, column temperature 100°C, filament temperature 250°C, 170 injector, split ratio 1:65. 12.5 m x 0.20 mm SE-5 column, column flow 0.79 ml/min. 9.45 ug hexadecane injected.

linear dynamic range is over 6 orders of magnitude, and the FID is flow independent. One can approximate the system variance (σ sys) to be:

$$\sigma^2 sys. = \sigma^2 FID = \sigma^2 inj. + \sigma^2 col. + \sigma^2 ct \qquad (17)$$

Subtraction of (17) from (16) yields the variance associated with the TCD, or:

$$\sigma^2 TCD = \sigma^2 total - \sigma^2 FID \qquad (18)$$

Determination of the TCD volumetric time constant by the method of statistical moments requires a minimum signal to noise ratio of 100, operation within the linear dynamic range of the TCD and FID being used, and sufficiently rapid data acquisition so as to preserve the integrity of the peak shape in the higher moments [13]. Experimentally the noise present for the TCD was too great to allow this method to be used with a hexadecane standard. Instead fast peaks were generated using a fast sampling valve (a modified Linear Dynamics Series 11, 2-way normally closed valve) to inject butane onto an uncoated 0.060mm x 2m column.

A plot of σ det versus flow $^{-1}$ reveals that the small volume TCD has an apparent volume of 48.8 $\mu\ell$. This extra column volume is due to other than solely turbulent mixing considerations. Sternberg [8] has described the possibility of symmetric band broadening, i.e. new spreading. This new spreading causes both symmetric spreading as well as large increases in the observed second moment. New spreading occurs when a sample pulse enters a wide diameter tube from a smaller diameter tube. Here the sample pulse will tend to retain the integrity of the peak shape on a column length basis within the detector; however, lateral diffusion leads to increases of the time based and volume based second moments (Figure 5).

At low flow rates mixing is very nearly complete, and the detector operates as an exponential dilution chamber due both to diffusion and turbulent mixing. As the flow rate is increased, the detector volumetric time constant decreases and the peak becomes more symmetric (see Figure 6). However, increasing the make-up flow rate any further, i.e. $F > 3\ V_{det}/\sigma_{FID}$, results in a symmetric broadening of the peak. Thus the TCD operates with both exponential mixing and new spreading convoluting the input peak shape.

Increasing the diameter of the connecting tubing to that of the TCD cavity diameter would eliminate this problem, at the expense of a vastly increased detector volume or laminar flow spreading. While new spreading cannot be eliminated, its chromatographic effects can be minimized by running the detector

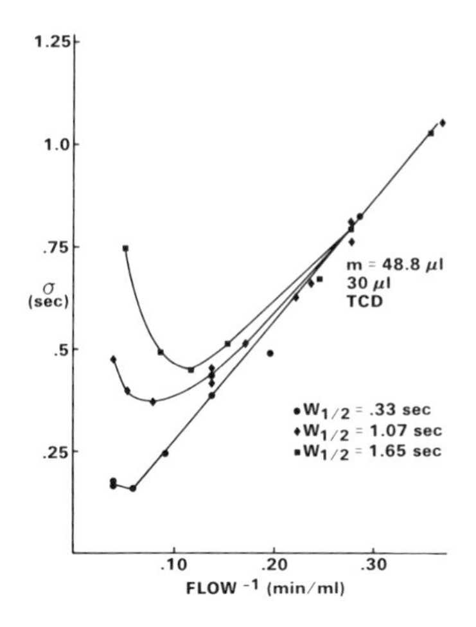

Figure 5. σ_{det} versus flow $^{-1}$. Fast sampling valve (50 ms) injected Butane onto an uncoated 0.060 mm x 2 M column. Detector temperature 120°C, filament temperature 180°C. Split ratio 1:300.

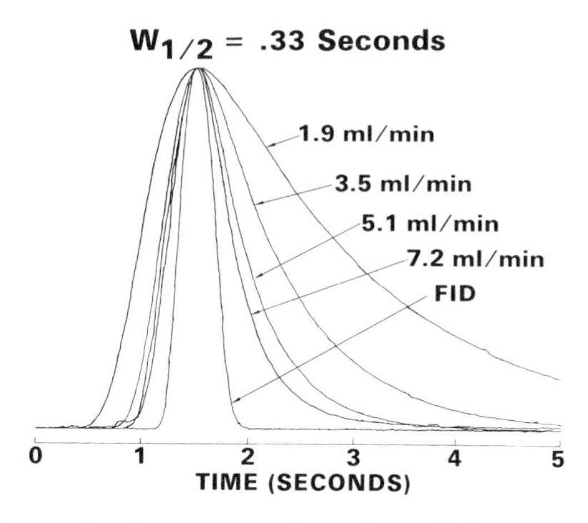

Figure 6. Peak shape as a function of detector flow rate. Conditions same as figure 5.

at its optimum flow rate for the expected peak width, or by compromising peak shape (i.e. fidelity) to gain in both sensitivity and to eliminate any excessive new spreading contributions to peak shape.

For packed columns, new spreading is a minor perturbation of the peak variance. Here the detector effectively operates as a plug filter on the Gaussian input function. Since $\sigma >> V_{det}/F_c$, the effects of new spreading is hidden in the uncertainty of the peak variance.

Using 0.20 mm x 12.5M SE-5 column, 1070 injector, column flow 0.78 ml/min, TCD temperature 170 ^{o}C, TCD filament temperature 250°C, the S/N varied from 50 at 2.90 ml/min to 5 at 33.5 ml/min for 2.50 μg hexadecane injected at a split ratio of 1:300. Increasing the sample injected to try and increase the S/N resulted in severe overloading of the 0.20 mm x 12.5M SE-54 column. This can be observed by examining the peak shape by expanding the time axis, and observing severe frontal distortion of the Gaussian peak shape. As sample concentration is increased the peak height remains constant and the peak width begins to dramatically increase. This serves to greatly decrease the resolution available using a capillary column. In effect, this negates some of the advantages that are inherent using capillary columns over packed columns regarding resolution, or detectivity. As the column sample capacity is exceeded, the peak width begins to increase dramatically (Figure 7).

Recently wide bore fused silica glass capillary columns have been generated which have very thick bonded phases (i.e. 5 μm) which are stable (14). These preparative columns have increased sample capacity (greater than 10 μg) while simultaneously maintaining high separation efficiencies. However, they also have the disadvantage of substantially increasing retention times. The use of one of these columns together with a 30 μℓ TCD would increase the dynamic range of the system a factor of 200. Thus for a 1 second peak half width, and a total detector flow of 9 ml/min, the dynamic range of the 30 μℓ TCD would be 4000. The chromatographic system would still be limited at high sample concentrations, but real quantitative work could be done.

A 5μℓ thermal conductivity cell has been discussed by Craven and Clouser (15,16) which is modulated at 10 cycles per second. Here the analytical and reference gas flows are alternately switched to flow through a single detection cell. An advantage of the modulated TCD is that the cell remains nearly constant for sequential pulses, and thus the modulated detection removes the low frequency components of the TCD signal. This effectively eliminates any problems that are associated with drifting baselines.

The modulated TCD retains the limitations imposed on conventional thermal conductivity detectors with respect to stability of the reference and analytical flows, detector wall temperature stability, and make-up flows being required to

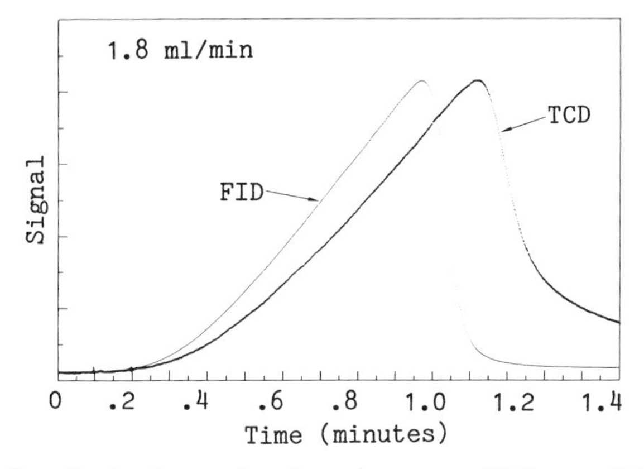

Figure 7. Peak shape for hexadecane. 12.5 m x 0.20 mm SE-5 column, 1070 injector, detector temperature 170°C, filament temperature 250°C, column flow 0.78 ml/min, split ratio 1:300, 46 ug hexadecane injected.

minimize band broadening at the column to detector union. In addition to the make-up flows being required to minimize peak broadening at the detector union, additional gas flow is required to efficiently sweep-out the detector volume and to effect the switching action. These additional flows through the cell are reflected as a decrease in the analyte concentration at the detector and thus reduce sensitivity. The modulation scheme itself results in a factor of 2 (at best) attenuation of the analytical signal.

Craven and Clouser have stated almost instantaneous baseline and sensitivity stabilization. Although the baseline will almost instantaneously stabilize, the sensitivity can not stabilize rapidly simply because the detector wall temperature will increase when the filament is turned on, and time is required for the wall to reach an equilibrium temperature. Indeed, most baseline drift for conventional TCD's is due primarily to temperature cycling of the detector wall and oxidation of the filaments. In the modulated mode, this drift is masked, but the sensitivity of the cell is always changing, as the detector wall temperature varies.

The modulation frequency of 10 Hz implies that the cell is being filled with the analytical flow for 1/20 second, and then during the next 1/20 second being purged by the reference flow. For a 5 µℓ TCD volume (assuming no diffusion into the reference flow lines of the analyte, no adsorption of the analyte on the cell walls, and totally reversible laminar flow through the cell), the minimum flow rate is 6.0 ml/min to totally fill or purge the cell. Any additional flow only serves to dilute the sample; however, any lower flow serves to not completely purge the analyte, resulting in lower sensitivity than can be theoretically attained. One should be aware that a significant component of the analytical and reference gas flows can consist of the modulator flow. The recommended modulator flow is 12.5 ml/min; however, in practice, even higher modulator flows are necessary for proper signal modulation. Thus simple measurement of the make-up flow and the column flow is not sufficient to describe the total gas flow across the filament.

The conventional TCD is configured with the filaments being connected to form a Wheatstone bridge. A property of the Wheatstone bridge is common mode rejection of the noise which is primarily due to the electronics (i.e. power supply stability and the amplifier circuit). The TCD noise spectrum resembles white (shot) noise rather than the 1/f (flicker) noise of ionization detectors. Modulation techniques for noise rejection of white noise is no better than a simple Wheatstone bridge.

In comparison, both the 30 µℓ and the 5 µℓ thermal conductivity detectors require good stability of the detector wall temperature, and matched analytical and reference flows. Although baseline drift is a real problem for the 30µℓ TCD systems which have air leaks, extremely poorly matched filaments, and poorly matched analytical and reference flows, a properly set-up capillary system tend to minimize these effects and show essentially no baseline drift for the 30μℓ TCD. The advantage of the 30μℓ TCD is that if baseline drift is present, then there is an indication of a problem, i.e. poor chromatographic technique. A steady baseline (with the Wheatstone bridge), reveals consistent sensitivity throughout the analysis, and for sequential analyses. This may or may not be the case with the modulated 5 μℓ TCD, since the modulation serves to remove slow drift components from the analytical signal, constant sensitivity from run to run (and thus reproducible quantitation) is not necessarily ensured.

Conclusions

(1) A small volume TCD still requires a reference flow to be well matched with the sample flow. For most wide bore WCOT columns, make-up gas is required to obtain the available resolution of the system. For SCOT columns this can be neglected without serious loss in efficiency. Regulation and balancing of the reference and sample flows are relatively easy when only isothermal operation is employed, but become very difficult when temperature programming is necessary.

(2) Linear dynamic range of the 30μℓ TCD is good when used with a packed column. When the 30μℓ TCD is coupled with WCOT columns, the limited sample capacity of the columns severely limits its analytical utility. While the dynamic range can be somewhat adjusted by judicious selection of the split ratio of a capillary injector, major and trace components can not be both simultaneously determined. The upper end of linear range is always limited by column capacity.

(3) Increasing the sensitivity of the detector, i.e. increasing ΔT of the TCD, gives a moderate (factor of 2 to 5) improvement in linear dynamic range. The limiting factors are the increased noise, and the degradation of the filament lifetime. High filament temperatures degrade the filament and allow rapid oxidation of the filaments if any air leaks are present.

(4) The 30 μℓ TCD in use with packed columns is of dubious usage. The 140 μℓ TCD has superior sensitivity at most packed column flow rates. Although the 30 μℓ TCD has marginal usage with wide bore (0.480mm) columns, neither the 30μℓ or the 140μℓ TCD have any practical usage with high resolution narrow bore columns (i.e. $d < .25$mm) columns. The 30 μℓ TCD may have some utility with SCOT columns with high stationary phase loading, but these were not investigated in this work. The 30μℓ TCD combined with the preparative

thick film (5 μm) bonded phase capillary columns would allow detection of up to 10 micrograms of a single component in the sample.

(5) While a 5μℓ modulated TCD is available for use in capillary chromatography, the 10 Hz modulation frequency limits the minimum peak width to be 1 second, and the modulation technique requires in large excess of 6 ml/min of flow through the detector cell. These factors severely limit the utility of the modulated cell for fast, high resolution capillary chromatography.

Literature Cited

1. D.J. David, Gas Chromatographic Detectors, Chapter 3. Wiley, New York (1974).
2. A.B. Littlewood, Gas Chromatagraphy, Chapter 9, Academic Press, New York (1970).
3. H. Purnell, Gas Chromatagraphy, Chapter 12, John Wiley & Sons, Inc. New York (1967).
4. S.D. Nogare, R.S. Juvet, Jr., Gas Liquid Chromatography, Chapter 10, Interscience Published, New York (1962).
5. A.E. Lawson, J.M. Miller, J. of Gas Chromatogr., 4 (1966) 273.
6. G. Wells, R. Simon, J. High Resol. Chromatogr. and Chromatogr Commun., 256 (1983) 1.
7. ASTM standards on Chromatography, Philadephia, Pa., ASTM E516-74 (1981) 661.
8. J.C. Sternberg, Adv. in Chromatogr., J.C. Giddings and R.A. Keller, Eds., Marcel Dekker, Inc., New York, Vol. 2 (1966) 205.
9. F.J. Yang, S.P. Cram, J. High Resol. Chromatogr. and Chromatogr. Commun., 2 (1979) 487.
10. C.H. Lochmuller, B.M. Gordon, J. Chromatogr. Sci., 15 (1977) 285.
11. C.H. Lochmuller, B.M. Gordon, J. Chromatogr. Sci., 16 (1978) 141.
12. C.H. Lochmuller, B.M. Gordon, J. Chromatogr. Sci., 16 (1978) 523.
13. S.N. Chester, S.P. Cram, Anal. Chem., 43 (1971) 1922.
14. K. Grob, G. Grob, J. of HRC & CC, 6, (1983), 133.
15. D.E. Clouser, J.S. Craven, U.S. Appl. 949312, 06 Oct. 1978.
16. D.E. Clouser, J.S. Craven, Analusis, 2, (1980), 3.

RECEIVED January 30, 1984

7

Advances in Liquid Chromatographic Selectivity

DAVID H. FREEMAN

Department of Chemistry, University of Maryland, College Park, MD 20742

Historical Development of Rapid HPLC Separations

In the past twelve years, HPLC, or high performance liquid chromatography, has been simplified. It has become easier to learn, and easier to apply to chemical separation problems. HPLC signifies the rapid achievement of high resolution separations.

The technique requires usually no more than a number of minutes for each compound separated. This aspect is shown in Figure 1 where the time for separating hemoglobin variants has been reduced from fourteen hours to 7 minutes. This trend, expressed in terms of the time needed to separate multi-component mixtures, has decreased steadily over the past forty years. A sampling of the separation literature illustrates this in Figure 2. The number of components separated per unit time when plotted as a logarithm is, approximately, a straight line function of the year reported.

The separation capabilities of current HPLC technology, as implied by Figure 2, did not arrive by a sudden jump into the "HPLC era". HPLC today is a cumulative advance, a result of progress made by a cast of many thousands who over the years have worked to expand the scope and depth of the field, and who have made their results known to others.

This growth has mushroomed into a cooperative social effect in the form of an informal network for information exchange. I estimate the availability of 100,000 HPLC instruments, at least the same number of liquid chromatographers, and perhaps several times that number of LC columns. Centers that spread information effectively and provide practical guidance include application laboratories, discussion groups, short courses and meeting symposia.

0097-6156/84/0250-0077$06.00/0

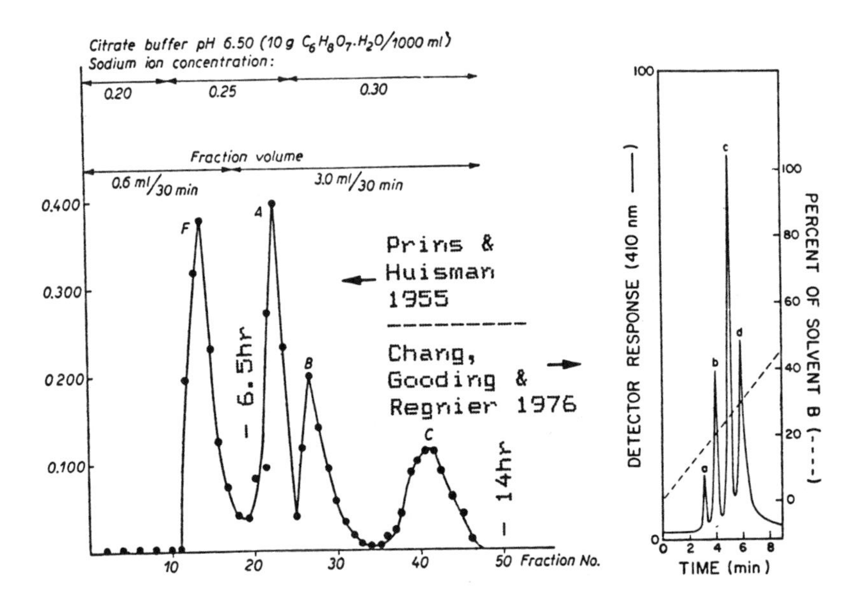

Figure 1. The separation of hemoglobin variants by classical LC(1) and by modern HPLC (2) techniques.

The number of trained chromatographers is expanding rapidly. Modern instrumentation is becoming more automated and easier to use. The gaining of practical experience is accelerated by contact with the available pool of other users. A portion of this is apparent in the published literature. To see the value of having acquired HPLC separation skills, one may examine the "help wanted" advertisements in Chemical and Engineering News.

The commercial popularization of HPLC was begun in the early 1970's at the same time that the Nobel laureate, the late R.B. Woodward, was pursuing the separation, and later the proof of purity, of the thermally labile cobester intermediates formed in the synthesis of vitamin B12.

One of Woodward's postdoctoral research associates, Helmut Hamburger, devoted a year's effort to trial and error applications of the use of silica gel adsorbent. The Harvard-MIT community knew of the difficulties being experienced. Professor Merrill of the MIT chemical engineering department, also on the board of directors of Waters Associates, communicated these difficulties to James Waters. His company, Waters Associates, had started an applications laboratory in 1967. Woodward's chemical insights were needed, as were Waters' technical insights, in order to achieve a proper harnessing of the speed and resolution of the new HPLC technology. It was decided that the problem was to be solved in Woodward's laboratory using the technology that Waters was eager to share.

The success of this venture is shown by the chromatogram in Figure 3. Subsequent work moved faster as HPLC was recognized as a more powerful basis for proving the purity of the isolated products. A significant precedent had been put to work: the willingness of industry to undertake the solving of difficult separation problems as a basis for subsequent sales of instrumentation and related supplies.

The rate of solving such separation problems is a less frequently discussed subject. An empirical approach is a tradition that even today is dominant in solving separation problems. Choosing or finding the right combination of liquid and stationary phases to obtain the needed chemical selectivity is usually an "educated" trial and error process. If analogous mixtures have been separated, that can provide a time-saving boost.

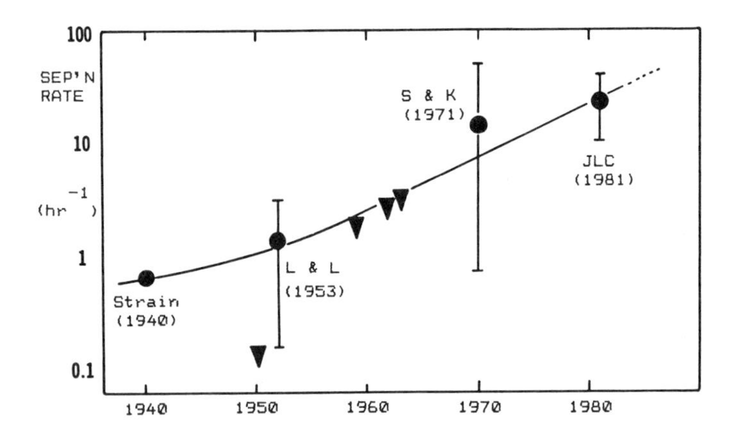

Figure 2. A sampling of separation rate, the number of peaks eluting per hour, against year of publication. Results are taken from Strain (3), L&L referring to Lederer and Lederer (4), S&K referring to Snyder and Kirkland (5), and JLC referring to the Journal of Liquid Chromatography (6). The inverted triangles refer to amino acid separations.

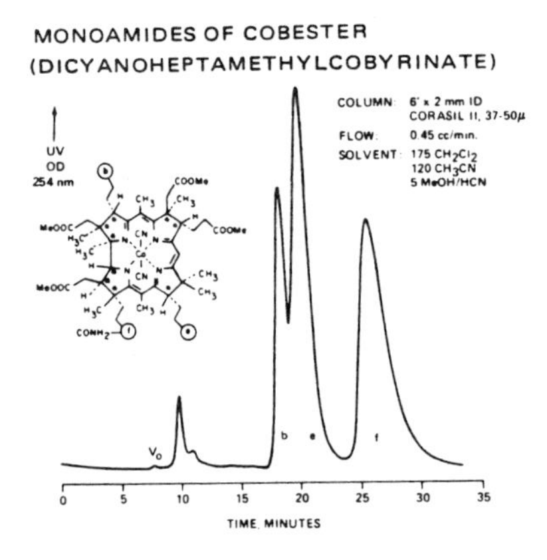

Figure 3. Isolation of cobester isomers where the f isomer is the key intermediate in the synthesis of vitamin B12. From the work of Woodward as reported in a commercial applications note (7).

An important development in HPLC has been the recent progress in systematizing column and solvent selectivity. The collection and improved accessibility of large arrays of empirical separation results are providing the major basis for faster problem solving.

The fast separation rates shown in Figures 2a and 3 do not convey the speed of obtaining the selectivity in the first place. If one were to add the time needed to find this selectivity (about one year) to the time axis shown in Figure 3, that would create a very different impression of whether this was an example of a "high speed" separation. Hence, to be realistic, the separation shown in Figure 3 required a time investment of a year plus half an hour. Since then, vast improvements have been made in the speed of finding chemical selectivity and these will be the theme in this discussion.

In its early beginnings, the trial and error approach involved numerous sorbents, and even more numerous combinations of liquids and additives. It follows that the complexity of LC problem solving has been at least in part related to the number of equally plausibly decisions needed in chosing an effective combination of liquid and sorbent. The likelihood of hitting the proper combination of liquid and sorbent by guessing was, and still is, quite small. In addition, the older adsorbents were not designed to function as such. Instead they were chosen because of their availability, their past applications history, and their estimated suitability based on considerations of their chemical structure. The occasional lack of reproducibility, low load capacity or sample size limitation, peak tailing, chemical reactivity, and uncertain solvent interactions slowed the problem solving process considerably. Before 1970, separation skills in LC were reserved for the relatively few practitioners who had gained expertese from relatively tedious experimentation.

Sorbent Selectivity

A crucial aspect of chromatography is the importance of minimizing the time needed for the solute to transit back and forth between the sorbent and liquid structures. This is necessary in order to obtain sharp peaks, as demonstrated by Martin and Synge in 1941 (8). This experimental principle was adapted to the ion exchange separation of amino acids

(15 min per component) in 1963 by Hamilton (9). Horvath (10) used impermeable beads as a base for an ion exchange resin film to separate nucleotides and nucleosides at 5 min per component.

These major strides in stationary phase design--maximizing the rate of mass transfer--are now added to a boosted sample load capacity. An optimal architecture, suited for purposes of speed and load capacity, is obtained when the sorptive functional groups are bonded to a highly porous substrate.

The sorbents most frequently chosen for current use are the porous silica derivatives. Most popular are the hydrophobic or "reverse phase" sorbents with bonded alkyl groups. These are ideally suited for the partly aqueous carrier phases that are suitable for separations related to the life sciences, an area that accounts for most of the current use of HPLC.

Many hundreds of chemical derivatives of porous silica have been prepared and tested (11). The ability to fabricate reproducibly a high performance LC column has considerable commercial value. As a result, column manufacturers are apt to be reluctant to convey the chemistry used to prepare the bonded functional group structures, but they often allow users to make educated guesses as to what those structures might be.

Porous silica microparticles less than 10 micrometers in size are used, among other purposes, to separate the least water soluble as well as the least interactive compounds. The use of basic functional groups (i.e. amine) on porous silica can be an appropriate choice for separations of at least slightly polar compounds that are hexane soluble, and even for aqueous separations of highly polar compounds, such as sugars in aqueous media. Ion exchange derivatives of silica compete with carefully prepared ion exchange resin (12), although the latter has one advantage in terms of its higher load capacity.

The only predictable selectivity in HPLC is that that provided by the non-interactive conditions that may be achieved with the porous adsorbents used in nonaqueous size exclusion chromatography. Predictability is based on a knowledge of the relative pore and solute sizes. The optimal size ratio for solute separation is four to one (13).

The most important stationary phase selections are usually aimed at the four major chromatographic categories of partition (normal phase on bonded amine or reversed phase with bonded alkyl), adsorption (i.e., silica), ion exchange and size exclusion.

There is a wide diversity of available trade names, synthetic nuances, or geometric ones such as column size and shape. However, over three-fourths of HPLC is done with chromatographic columns prepared as porous silica derivatives with bonded alkyl functional groups. Other bonded groups that are noteworthy include diol functionality for aqueous size exclusion and bonded counterions for ion-exchange. The result is an increasingly well focussed interest in relatively few, but widely used, column types. It is known that nominally identical column types do not exhibit identical separation behavior in practice. This non-uniformity has provoked user interest in the question of the extent to which sorptive reproducibility occurs from column to column and, similarly, among different manufacturers. Researchers, as well as applications specialists, must have well defined sorptive structures if their results are to be truly intercomparable.

One basis for irreproducible column performance among the bonded phases comes from the difficulty in removing the silanol groups from the silica substrate. In some applications the groups provide needed additional interactions while in other cases this is unwanted. Currently it is difficult to measure the silanol concentration definitively, and it remains a synthetic challenge to completely remove the silanol groups through end-capping reactions.

SOLVENT SELECTIVITY

Improvements in stationary phase design have advanced to a more coherent technology for achieving, at least in practical terms, well defined sorptive effects. Equally important, similar progress has been made toward improved practical understanding of liquid phase compositions needed to achieve chemical selectivity. Preliminary solvent selection has been reduced to the use of solvent triads, one for aqueous and another for non-aqueous systems (14). Thus, aqueous mixtures for reversed phase HPLC, or RPLC, are prepared with methanol, acetonitrile, and/or tetrahydrofuran as liquid phase modifiers. Increasing the water content in RPLC intensifies the lipophilic interactions between sample and stationary phase.

In normal phase HPLC, or NPLC, hexane is a widely used diluent and the common modifiers are one or a combination of dichloromethane, chloroform and ether. Remember, bonded amine derivatives of silica are also used for normal phase separation of sugars and other highly polar compounds.

For simple mixtures selectivity can often be found quickly. If the sample mixture is at least sparingly water soluble, RPLC may be the first choice. If the sample is water insoluble, then NPLC is preferred. Once the stationary phase is chosen, testing within the solvent modifier triad is then usually done by trial and error. Such experiments can be carried out by automated instrumentation programmed for repetitive scouting experiments.

Those who deal routinely with the widely varied separation problems that show up in applications laboratories share the opinion that the easy problems in HPLC have already been solved. Application chemists estimate informally that simple solvent blending with relatively few additives will provide the selectivity needed to separate 90% of the routinely encountered mixtures. User support is a key marketing activity that includes the provision of estimated separation conditions or enhancements, recommended columns, solvents, additives and other tips or recommendations. This may be provided, for example, by an applications laboratory in response to a telephone inquiry.

Separation problems become substantially more difficult as the number of components increases much above ten. Such complexity is often characteristic of environmental and many biological samples. Ionic interactions and ion exchange, for example, may offer potentially unlimited selectivity but the conditions for optimal selectivity are correspondingly more difficult to find.

ROLE OF ADDITIVES

The growing separation capabilities of HPLC include an expanding catalog of liquid phase additives. Such compounds are placed in the carrier in order to stabilize or augment the chemical selectivity. Consider the following examples: (a) Reverse phase partitioning of carboxylic acids or amine compounds is more definitive when the compounds are kept in their less polar state; i.e., when the carboxyl groups are protonated (non-ionic and therefore less polar) by the presence of acid. Similar effects are obtained at higher pH with the use of basic media for converting ammonium to amine. Such conditions, known as ion suppression, are generated in the carrier phase. (One limitation is the increasing solubility of the silica substrate above pH 8.) (b) Excessive electrostatic interactions of metal ions are suppressed by complexing anions, such as the use of gamma-hydroxybuterate for the cation exchange of rare earths (15). (c) Unwanted chelation effects take place between certain antibiotics and the stainless steel in LC systems. This effect can be swamped out by adding EDTA to the carrier.

CARRIER INDUCED SELECTIVITY

Experiments show that selectivity can be introduced, or induced, on the basis of the carrier composition alone. To illustrate, ion exchange has been regarded as the main separation process that results when electrolytes are chromatographed on a stationary phase having bound counterions. It is now recognized that the same ion exchange mechanism can be achieved in the absence of fixed charge sites. This is achieved, for example, by addition to the carrier liquid of hydrophobic ions under reverse phase conditions. The explanation is that the hydrophobic ends of the ions are sorbed by the lipophilic sorbent. As a practical example, the counterions needed for cation exchange can be obtained from pentyl sulfonate salt added to the carrier. Similarly, anion exchange is obtained in the presence of carrier with octyl ammonium salt. Such effects are turned off by stopping the additive feed to the carrier (16).

The use of the preceding carrier induced ion exchange effect is notably complicated because the underlying reverse phase partitioning effect is still operational. This occurs with ion exchange resins as well. The existence in a single stationary phase of multiple chromatographic mechanisms has the effect of expanding the basis for selectivity, but it also adds to the difficulty of chromatographic optimization.

When utilizing the ion exchange mechanism the degree of solute ionization is controlled by pH, and the partitioning effect is responsive to carrier solvent blending. These are independent variables. It is important to know that HPLC offers options to control selectivity though the mobile phase in order to induce ion exchange, to alter the ionic charge state of the sample or of the stationary phase, or to remove either or both of these charge effects by controlling the carrier feed composition. An important spin-off from these advances is they are leading toward improved chemical understanding of the behavior and properties of the liquid phase.

An impressive separation of optically active amino acids has been demonstrated through the use of chiral reagents that induce the ligand exchange mechanism (17). The result is illustrated in Figure 4. The D,L elution order of the compounds depends upon whether a D or L chiral reagent is used, and the chiral separation terminates if a racemic reagent is used.

While only a few illustrations of carrier phase control of HPLC selectivity have been given, it should be recognized the harnessing of liquid phase composition to control HPLC selectivity is providing a major corridor for achieving separations in an increasingly systematic manner.

ULTRA-RESOLUTION

A systematic basis for the combining of independent selectivity mechanisms can provide a major boost to the overall selectivity. The overall effect is multiplicative based on the separating power, or peak capacity, of each of the steps. Either implicitly or explicitly, this is the widely used basis for multi-step separation schemes.

The serial implementation of multiple origins of selectivity is the most practical approach at present. It has been adapted for columnar LC using a "heart cutting" technique (18) introduced in gas chromatography.

The use of a number, N, of different chromatographic mechanisms in sequence is known as column switching. Assume each mechanism has the same effective peak capacity. In its simplest form, the use of such a chromatographic sequence expands the

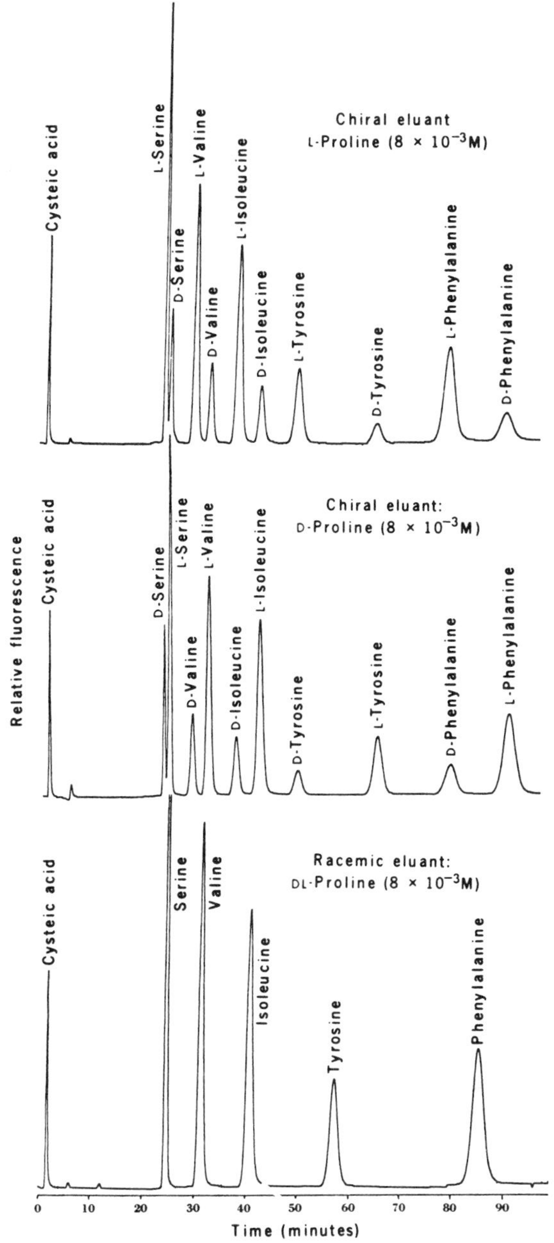

Figure 4. Separation by HPLC of the D,L amino acid enantiomers using a chiral reagent. The D,L order is obtained when L-Proline is added to the carrier in the presence of cupric sulfate. The enantiomer resolution is reversed using D-Proline. No optical resolution is obtained when the carrier chiralty is removed, i.e., with the use of the racemic D,L eluant (17).

overall selectivity of the LC system by the Nth power of that obtained from a single selectivity mechanism. To illustrate, the serial combination of three different column mechanisms, each having a value of ten as its effective peak capacity, will be expected to be able to separate a thousand components.

Experimentally, two column switching can be demonstrated using apparatus no more elaborate than a conventional six-port valve. Column switching can create tremendous separating power, but it is a requirement that each one in the sequence of selectivity mechanisms not be redundant. On this basis it was calculated that a trillion compounds could be separated with available column switching technology (19).

The isolation of specific compounds using ultra-selective column switching HPLC is different in concept from the separation of all compounds in a sample mixture in a single chromatogram. The latter is a typical goal in capillary gas chromatography. Column switching should be viewed instead as a series of group separations where the realization of individual compound separation, i.e. the final sub-classification or sub-group, would be the last step in the chromatography. The limitations of the single chromatogram approach are discussed by Giddings elsewhere in this symposium (20).

If the number of independent contributions to selectivity from the carrier is considered in combination with the four cited stationary phase categories, the peak capacity of an HPLC system, referring to the maximum number of separable compounds, becomes exceedingly large, and it may well be unlimited. This reasoning points to the advantage of combining short columns in an ultra-high selectivity scheme, as opposed to the use of a single ultra-high resolution column.

Recent breakthroughs over the past two or three years in protein separations by HPLC can be summarized on an approximate numerical basis using the foregoing considerations based on peak capacity. A rough estimate of presently available peak capacity for proteins can be estimated from the available peak capacities: from aqueous size exclusion (approximately 5), aqueous partition (100), and ion exchange (200). The estimate of 200 for the latter includes estimated selectivity contributions due to pH, salt, and organic modifier. The cumulative number (the product of individual peak capacities) of HPLC separable proteins is thus estimated at over 100,000.

HPLC has not completed its growth nor is it near approaching its limit. To be sure, simple separation problems are now greeted by an excess of separation capability. The separation of complex mixtures and the streamlining of multiple selectivity applications is of growing interest. Only partly developed is the need for sensitive and selective detectors. This area is felt by many to be the weak link in HPLC where there is still much room and need for advancement.

Literature Cited

1. Prins, H.K. and Huisman, T.H.J., Nature 175, 903 (1950).

2. Chang, S.H., Gooding, K.M. and Regnier, F.E., J. Chromatogr., 42, 355 (1976).

3. Strain, H., "Chromatographic Adsorption Analysis," Revised. Interscience Pub., Inc. New York, 1945.

4. Lederer, E. and Lederer, M., "Chromatography," Elsevier, Amsterdam, 1957.

5. Snyder, L.R. and Kirkland, J.J., "Introduction to Liquid Chromatography," Wiley-Interscience, New York, 1979.

6. J. Liquid Chromatography 4, 1-2290 (1981).

7. Applications Note AN 120, Waters Associates, Milford, MA (1972).

8. Martin, A.J.P. and Synge, R.L.M., Biochem. J. 35, 1358 (1941).

9. Hamilton, P.B., in "Handbook of Biochemistry, Selected Data for Molecular Biology" Chemical Rubber Co., Cleveland, 1968, p. B-47.

10. Horvath, C.G., Preiss, B.A. and Lipsky, S.R., Anal. Chem. 39, 1422-1428 (1967).

11. Unger, K.K., "Porous Silica," Elsevier, Amsterdam, 1979, Chap. 3 (249 references).

12. Elchuk, S. and Cassidy, R.M., Anal. Chem. 51, 1434 (1979).

13. Freeman, D.H. and Poinescu, I.C., Anal. Chem. 49, 1183-1188 (1977).

14. Snyder, L.R., "Gradient Elution" in "High Performance Liquid Chromatography," C. Horvath, Ed. Academic Press, New York, 1980.

15. Blasius, E., Augustin, H., Hausler, H., Janzen, K.-P. and Wagner, H., in Heftman, E., "Chromatography, a Handbook of Chromatographic and Electrophoretic Methods," 3rd Ed., Van Nostrand, Reinhold, New York, 1975, pp. 847-853 and 860.

16. Skelly, N.E., Anal. Chem. 54, 712 (1982).

17. Hare, P.E. and Gil-Av, E., Science 204, 1226-1228 (1979).

18. Deans, D.R., Chromatographia 1, 12-22 (1968).

19. Freeman, D.H., Anal. Chem. 53, 2-5 (1981).

20. Giddings, J.C., Davis, J.M., Schure, M.R.; Chapter 2 in this book.

RECEIVED March 5, 1984

8

Narrow-Bore, Microparticle-Packed Column High Performance Liquid Chromatography

Utilization and Constraints

FRANK J. YANG

Walnut Creek Division, Varian Instrument Group, Walnut Creek, CA 94598

An overview of microbore column technology is given. The potential and the constraints of microbore columns are discussed. The driving forces in microbore column HPLC are (1) the compatibility of the flow rates with mass spectrometer and flame-based detectors; (2) a total saving of up to 99.9% solvent consumption; (3) the high resolving power of the long fused-silica columns packed with small particles; and (4) the opportunities in new detector development and new applications. To utilize the available potential of microbore packed columns, the liquid chromatograph including injectors, pumps, and detectors must be improved to match the chromatographic requirements. Modifications of existing commercial liquid chromatograph for the practice of microbore column HPLC are also described.

The driving forces for the rapid development and growth in microbore column HPLC are (1) savings in solvent consumption: a total saving of up to 99.9% can be achieved when narrow-bore microparticle packed columns or open-tubular micro-capillary columns are used; (2) the high separation power using long column and small particles (e.g., 3 μm); (3) the compatibility of the column eluent flow rates with a mass spectrometer and flame based detectors; and (4) opportunities in new detector development, e.g., a laser based spectroscopic detector. Microbore column HPLC, however, requires substantial improvement of the conventional liquid chromatograph to meet the operational requirements and chromatographic performance of microbore columns.

It is the purpose of this manuscript to compare four types of microbore columns reported in the literature. The utilization of the narrow-bore packed column HPLC in terms of instrumentation requirements for injection valves, pumping systems, and detectors are also discussed.

0097-6156/84/0250-0091$06.00/0

Columns

Since the introduction of open-tubular columns for gas chromatography by Golay[1] in 1958, developments in open-tubular column gas chromatographic technology and techniques have made open-tubular column GC a very important tool for achieving high separation power and speed for complex sample analysis. In liquid chromatography, attempts to achieve high resolution and high speed LC analysis have been made by many researchers in the development of microbore column HPLC. Table I shows the operating parameters of four types of microbore columns reported in the literature. A brief comparison of these four types of microbore columns are given below.

Open-Tubular Columns

Open-tubular columns for liquid chromatography were investigated by Ishii and associates,[2-4] Yang,[5,6] and Tijsen et. al.[7] Columns reported are 10-300 μm I.D. and 0.5-35 m in length. LC stationary phases were either physically coated or chemically bonded on the inner wall of the column. The typical mobile solvent flow rates are normally ranged from 0.01 to 1 μℓ/min. The theoretical performance potential and practical constraints of open-tubular column LC have been discussed by Yang,[6] Tijssen,[7] Pretorius and Smuts,[8] Guiochon,[9] and Halasz et. al.[10] Theoretically, open-tubular column could achieve state-of-the-art packed column performance in terms of speed of analysis, column efficiency and peak capacity, if the open-tubular column I.D. could be reduced to approach the particle diameter of the state-of-the-art LC packed columns. For example, open- tubular columns with I.D. less than 7.5 μm have better resolving power than that obtainable by using the state-of-the-art 5 μm particles packed columns in a given analysis time.

Many reports[6,9,11] have been given dealing with the constraints of open-tubular column LC instrumentation. There is no doubt that open-tubular column LC poses the most stringent requirements on the pump (e.g., 0.01 μℓ/min controlled flow rate), interface tubing and connectors (e.g., ≤1 nℓ). The theoretical performance potential of open-tubular column LC is thus difficult to achieve. Although splitter injector[5,12] and on-column optical detection[13] eliminate the need for column to injector and detector interfacing, and meet sub-nanoliter detector cell volume requirement, a pump which can control flow rates in sub-μℓ/min range is extremely difficult to design, based on present hydraulic pump technology. With the use of a constant pressure syringe pump, sub-μℓ/min flow rates can be obtained; however, good flow rate stability and resettability at the low flow rate ranges are difficult to obtain.

Table I. Characteristics and Operational Parameters for Microbore Columns.

	COLUMN TYPES	MATERIAL	COLUMN I.D. m	PARTICLE SIZE μm	COLUMN LENGTH m	FLOW RATE μℓ/min
	Open-Tubular Column	Soft Glass and Fused-Silica Glass	10-300	--	0.5-35	0.01-1
MICROBORE PACKED COLUMN	Microcapillary Packed Column	Soft Glass and Pyrex Glass	50-200	10-100	10-60	0.01-5
	Small-Bore Packed Column	Stainless Steel and Teflon	500-1000	5-20	0.1-1	30-100
	Narrow-Bore Packed Column	Pyrex Glass PTFE Fused-Silica Glass Stainless Steel	100-500	3-10	0.1-2	0.1-20

Microcapillary Packed Columns

Novotny et. al.[14,15] reported microcapillary packed columns of 50-200 μm I.D. in 10-60 m length. Microcapillary packed columns were, in general, packed with high temperature resistant particles which have a diameter of about one third of the column diameter. The columns have low flow resistance, thus long column length can be used. Because of the large particle size (e.g., 30 μm) used, the performance of the microcapillary packed column is poorer than the conventional packed column of 3 and 5 μm particles. Although long microcapillary column is used in the separation of relatively complex samples, the increased resolving power has to be paid for in long analysis time and loss in peak detection. Microcapillary packed column LC can be used in conjunction with commercially available micro-volume valve injectors (e.g., $\leq$0.2 μℓ valves) and micro-flow cell detectors (e.g., $\leq$0.5μℓ UV detectors), if a long column length is used. In terms of flow rate and pump requirements, packed microcapillary column LC has about the same operational flow rate ranges thus the same difficulties in flow rate control are resulted as that for the open-tubular column LC. A constant pressure syringe pump is normally used in microcapillary packed column LC.

Small-Bore Packed Columns

Scott and Kucera[16-22] developed small-bore packed columns of inner diameter between 0.5 and 1 mm. Column length up to 1 meter can be efficiently packed with 30 μm particles. For small particles, only a short length column can be efficiently packed. A 30 cm x 1 mm I.D. column packed with 5 μm C_{18} particles achieves 30,000 plates.[23] The advantages of small-bore packed columns are (1) saving in solvent and packing material consumption (e.g., 20 to 30 times compared to columns of 4 mm I.D. operated at 1 mℓ/min flow rate); and (2) it may be directly interfaced into a mass spectrometer if column flow rate below 40 μℓ/min is used. In addition, small-bore packed column LC may be practiced with conventional instrumentation when a long column length is used. Flow rates in the range of 10-100 μℓ/min can be operated by some commercially available pumps or by modification of some conventional pumps. Conventional injectors and detectors with 0.1-0.5μℓ volume can be satisfactorily used with long columns. For high speed separation using short column length, however, injectors and detectors of less than 0.1μℓ may be required for peaks eluted with small capacity ratios.

Narrow-Bore Packed Column

Narrow-bore packed-column HPLC, as reported by Yang[12] utilized fused silica columns of $\leq$500 μm I.D. packed with 3-, 5- and 10-μm particles. A proprietary packing technique yields high efficiency 3 μm particle packed columns up to 2 m in length. For example, a 330 μm x 1 m, 3 μm C_{18} reverse-phase packed column[12] achieves 110,000 theoretical plates for pyrene at k′=6 in the linear flow velocity range between 0.5 and 1.5 mm/sec.[12] A 2 m x 0.3 mm I.D. 3 μ C_{18} column[24] produced a column efficiency of 144,000 plates for pyrene at k′=10. Short fused silica narrow-bore packed columns (10-20 cm) have also been reported by Ishii[47-49] using low pressure packing technique. The utilization of a single long narrow-bore column to achieve high separation power is advantageous relative to connecting short columns in series. It is easy to handle and does not risk band broadening due to the void space in the connecter unions joining the short columns together. Figure 1 shows the application of the high plate count available in such a long fused-silica 3 μm C_{18} packed column (1 m x 0.32 mm I.D. 92,000 plates measured for dibenzo(a,b)anthracene in the isocratic separation of an Environmental Protection Agency (EPA) priority pollutant mixture of sixteen PAH components.

In addition to the high separation efficiency and resolving power of the fused-silica narrow-bore microparticle packed columns, several other practical advantages are cited below:

1. The low mobile phase flow rates (0.1 to 20 μℓ /min) used in narrow-bore packed columns allow not only between 50 and 500 times solvent saving compared with columns operated at 1 mℓ/min, but also provide the potential for direct interfacing to mass spectrometers[25-33] and flame-based detectors.[34,35]
2. The high mechanical strength of the fused-silica tubing allows an inlet pressure range of up to 800 atm. Long length high pressure packed columns can be prepared and utilized. Columns can be coiled, simplifying the design of the column oven.
3. Good optical transparency of the fused silica glass allows the use of the column end as the flow cell for on-column UV absorbance or fluorescence detection.[24] It also allows visual observation of the packed bed for packing uniformity.
4. Good flexibility and small column O.D. of the fused silica columns allow easy connection to injector and detector. Multiple columns can be connected to an injector (i.e., also one pumping system), providing the potential for increased instrument throughput and an increase in analytical information obtained in comparison to that using a single column LC.

Narrow-Bore Packed Column HPLC Instrumentation

To utilize the available potential of narrow-bore packed columns, the liquid chromatograph including injector, hydraulic pumping system, and detectors must have a matching level of performance.

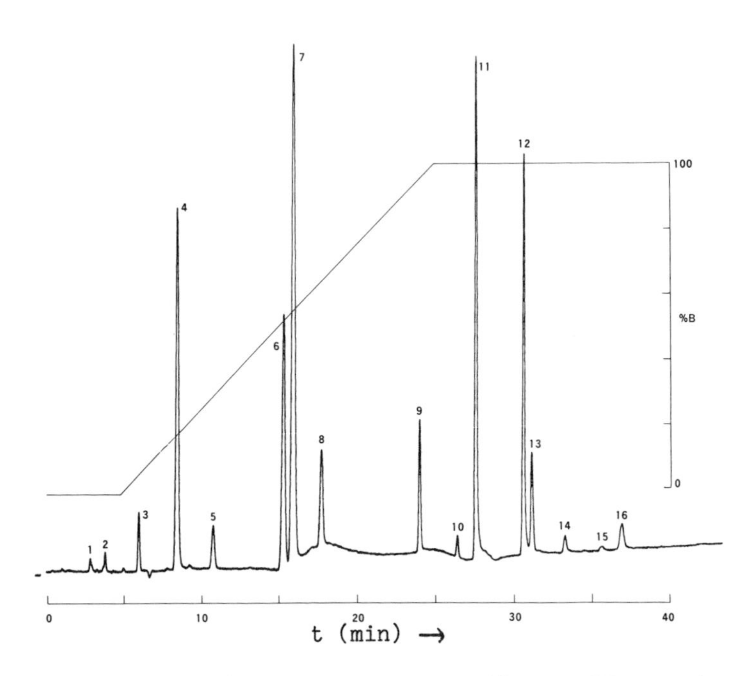

Figure 1. Separation of an EPA priority pollutant PNA sample using a 1 m x 320 µm, 3 µm C_{18} reverse phase fused-silica column. Mobile solvent was 70% ACN:H_2O at 0.95 µℓ/min. On-column UV detection at 254 nm and 0.01 A was used. Peak identifications are 1, naphthalene; 2, acenaphthalene; 3, acenaphthene; 4, fluorene; 5, phenanthene; 6, anthracene; 7, fluoranthene; 8, pyrene; 9, benzo(a)anthracene; 10, chrysene; 11, benzo(b)fluoranthene; 12, benzo(k)fluoranthene; 13, benzo(a)pyrene; 14, dibenzo-(a,h)anthracene; 15, benzo(ghi)perylene.

There is no doubt that narrow-bore packed columns pose very stringent requirements on the design of the liquid chromatograph. Modifications of existing commercial equipment for the practice of narrow-bore packed column HPLC are given below:

Injector

For high pressure narrow-bore packed column HPLC, a splitter injector utilizing a conventional internal loop rotary valve was reported by Yang[12] and is shown in Figure 2. The valve injector splitter is compatible with operating pressures of up to 500 atm. The splitter injector allows injections of small sample volume into the column and it also minimizes the injector extra column dead volume effect. At high split ratios, an improvement of more than 50% in degree of peak spreading measured for the direct injection can be obtained. However, high split ratios may not be compatible with trace analysis. Splitter injector quantitative reproducibility was reported to be better than 2.1% relative standard deviation for repetitive injections.[24]

Direct injection of small sample volume using a high pressure microvolume valve is desirable in microbore column HPLC. A special microvolume valve injector was developed by Takeuchi and Ishii[36]. A peak height reproducibility of 0.6% standard deviation was measured for repetitive injections of 20 nℓ PAH samples. A direct rotary valve injector was also employed by Scott et. al.[16] in their small-bore packed column HPLC work. Recent developments in valve technology have produced a commercially available 70 nℓ internal loop valve from Valco (Valco Instruments Co., Houston, Texas). With the addition of an electrically switched actuator, a sample volume as small as 20 nℓ may be injected. The extra column peak spreading contribution of the 70 nℓ valve was measured in our laboratories to be negligible for flow rate ranges between 3 and 5 μℓ/min.

Direct injection of large sample volume (>1 μℓ) into narrow-bore packed columns is allowed if a solute on-column focussing technique[12,24] is applied. In principle, a weak solvent modifier can either be present in the sample solution or be introduced just before the injection of the sample. An example of the on-column focussing effect on peak resolution is shown in Figure 3. Figure 3-a shows a chromatogram of four thiol compounds eluted from a 50 cm x 250 μm, 5 μm C_{18} reversed phase column using 100% methanol as solvent for both mobile phase and sample solution. Figure 3-b shows the improvement in peak resolution and detection obtained under the same conditions except that the injected sample solution contains 13% water in methanol. The effect of the weak solvent modifier depends upon the relative solvent strength of the solvent modifier to the mobile solvent and the amount relative to the column volume. In general, a weak

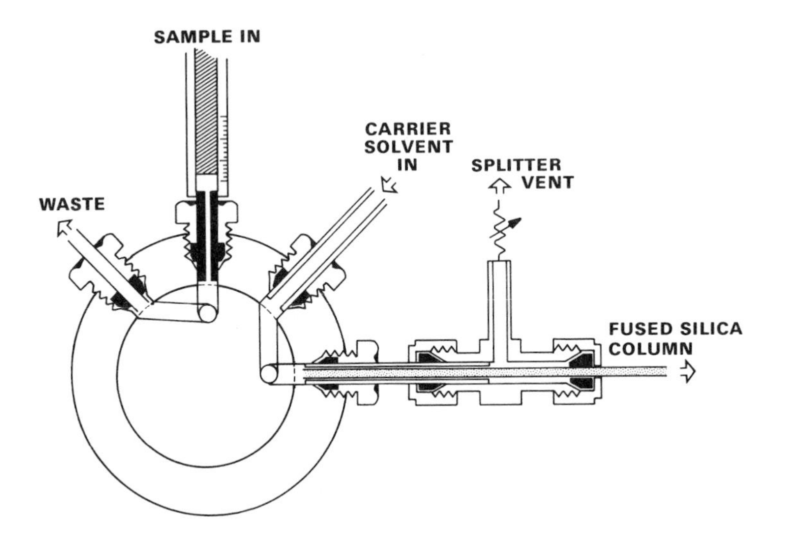

Figure 2. The modified Valco 0.5 μℓ internal loop sampling valve for split injection in narrow-bore packed column HPLC (Reproduced with permission from Ref. 12, copyright Elsevier Sci. Publ. Co.)

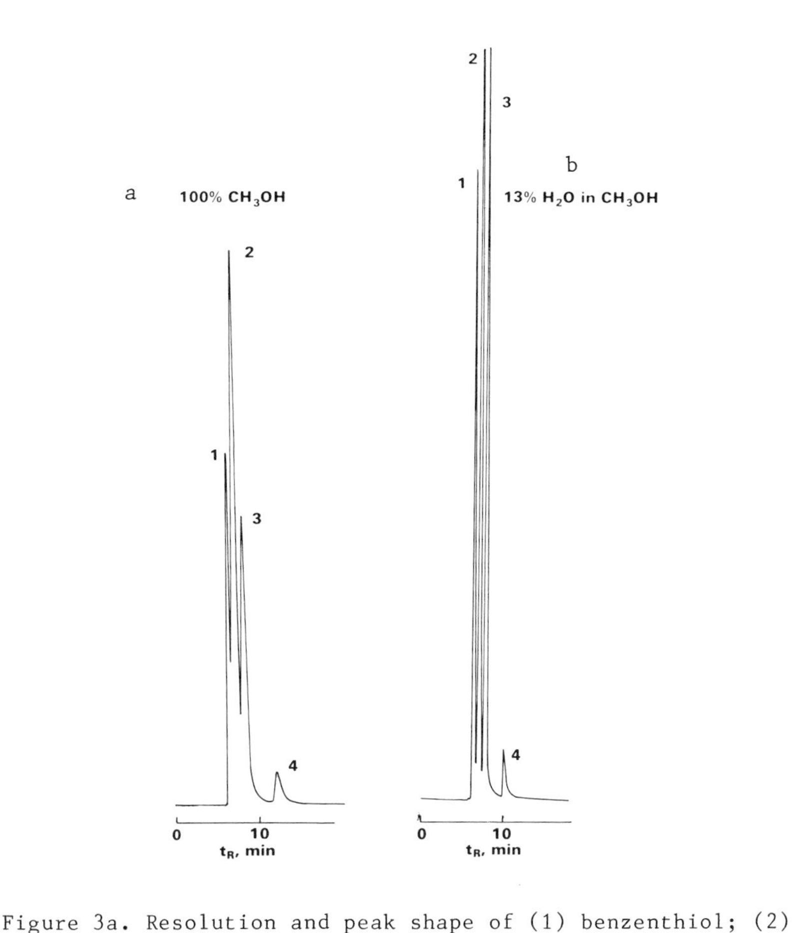

Figure 3a. Resolution and peak shape of (1) benzenthiol; (2) hexanethiol; (3) octanethiol and (4) dodecanethiol eluted from a 50 cm x 250 μm, 5 μm C_{18} reverse phase fused-silica column. Methanol was used as solvent for both mobile and sample solution. Sample volume was 0.6 μℓ.

Figure 3b. Improved peak resolution and detection by using 13% H_2O in the 0.6 μℓ sample solution. All experimental conditions were identical to that for Figure 3a. (Reproduced with permission from Ref. 24, copyright Preston Publ. Inc.)

solvent modifier volume which is more than 1% of the column volume can produce the effect of on-column solute focussing in narrow-bore packed column HPLC.

Hydraulic Systems for Isocratic and Gradient Microbore Column HPLC

The design of hydraulic systems for narrow-bore packed column HPLC is particularly challenging because of the low flow rate used in narrow-bore packed columns. The potential performance of narrow-bore packed columns is greatly reduced by the constraint of operable constant low flow rates and the operable pressure limit.

A high pressure syringe pump operated under constant pressure mode is often used for low flow rate operation in narrow-bore packed column HPLC. Flow rates as low as 0.05 $\mu\ell$/min can be obtained. The drawback of constant pressure operation is the potential flow rate variation due to a flow resistance change in the column.

In gradient elution narrow-bore packed column LC, both step gradient(12,37) and continuous gradient(16,20,38,39) systems were reported. In step gradient operation, a strong solvent stored in an injection loop is introduced into the microbore column at a selected time using an inline injector switching valve. A multi-port valve allows the application of multiple step gradient elution. The advantages of the step gradient technique are that only a single pump is needed to achieve the effect of gradient elution and the gradient delay time is minimum.

A linear gradient has advantages in terms of automation and ease in practice. Takeuchi and Ishii(38) also described a continuous gradient system utilizing a syringe pump and a syringe type mixing vessel. A single reciprocating hydraulic pump modified for linear gradient elution in microbore column HPLC has recently been reported by van der Wal and Yang(39) and is shown in Figure 4. The use of such a gradient system in conjunction with a narrow-bore fused-silica microparticle packed column at a flow rate between 3 and 5 $\mu\ell$/min showed a retention time and peak area reproducibility comparable to that obtained by a conventional HPLC system.

Detector

Because of the small elution peak volume obtained from narrow-bore packed columns, conventional concentration dependent detectors such as UV-visible absorbance, fluorometric, and electrochemical detectors must be purged with makeup solvent or miniaturized to allow minimum extra-column contribution to peak spreading. A cell volume $\leq 0.1 \mu\ell$ is desirable for narrow-bore packed columns of I.D. Ishii et. al.(38) reported the reduction of an UV detector cell volume to 0.4$\mu\ell$ by using a quartz tube of

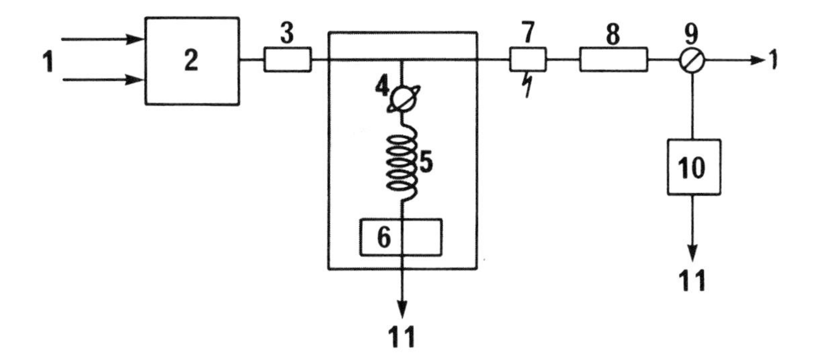

Figure 4. Schematic of microbore column HPLC hydraulic system: 1) solvent reservoir; 2) Varian Model 5000 pump; 3) static mixer; 4) injection valve; 5) narrow-bore fused-silica column; 6) detector; 7) pressure transducer; 8) pulse damper; 9) purge valve; 10) flow controlling device; 11) waste. (Reproduced with permission from Ref. 39, copyright Dr. Alfred Heuthig Publishers.)

1.5 cm x 0.18 mm I.D. for the flow cell. Yang[13] proposed the use of fused-silica tubing for narrow-bore packed columns, which allows on-column UV absorbance and fluorescence detection. The on-column detection technique as shown in Figure 5 provides the simplest way of eliminating column-detector interface tubing and peak broadening due to detector extra-column peak spreading effects. Figure 6 shows the separation of an EPA priority pollutant PAH sample eluted from a 66 cm x 0.25 mm I.D. fused-silica column with 5 μm Vydac 201 TP particles, using a linear gradient and on-column fluorescence detection.

The miniaturization of the spectroscopic absorbance and fluorescence detector flow cells is important from the view of minimizing post column peak spreading. However, the reduction in flow cell volume and optical pathlength significantly reduce the response of the fluorescence and absorbance detectors respectively. This can be balanced by a reduction in noise. Laser-based spectroscopic detectors are thus ideal for the detection of microbore column eluent. The laser, because of its high power levels relative to conventional sources, offers potentially lower noise levels. The laser's narrower output beam allows greater ease in focussing through micro flow cells. Folestad et. al.[40] reported a laser induced fluorescence detector for microbore column eluent detection. A flow cell volume of 65 nℓ was reported. A detection limit of 20 fg for fluorathene was measured under the conditions used for the free falling jet eluent. Potential applications of laser-based spectroscopic detectors such as a laser-induced thermal lens calorimetry detector, refractive index and absorption detector, etc., for bore micro-HPLC eluent detection have also been suggested.[41-44]

The use of FT-IR as a narrow-bore packed column HPLC detector has also been investigated by Jinno et. al.[45.46] Direct interfacing of LC-FT-IR is performed via a flow cell technique. To avoid the interferences from mobile phase absorbance in the mid-IR region, a compatible (transparent) mobile phase should be used. A "Buffer Memory" technique for indirect combinations of microbore column and FT-IR was also reported by Jinno et. al.[46] The buffer memory technique was reported to have advantages over conventional direct LC-FT-IR in achieving a continuous chromatogram and in preventing mobile phase interference.

The technique of direct LC-MS interfacing has been described in detail by Henion[25-28] and Tsuge and associates.[30-33] The direct LC-MS interface can provide a 1-50 ng full scan chemical ionization (CI) mass spectrum of underivatized and thermally labile drugs.[27] The low flow rate of narrow-bore column HPLC may facilitate the application of deuterated solvents for the characterization of ion molecule reactions and trace level impurities. The chromatographic performance of a direct LC-MS interface system was studied by Schaffer et. al.[29] They reported that reproducible spectra could be obtained only when

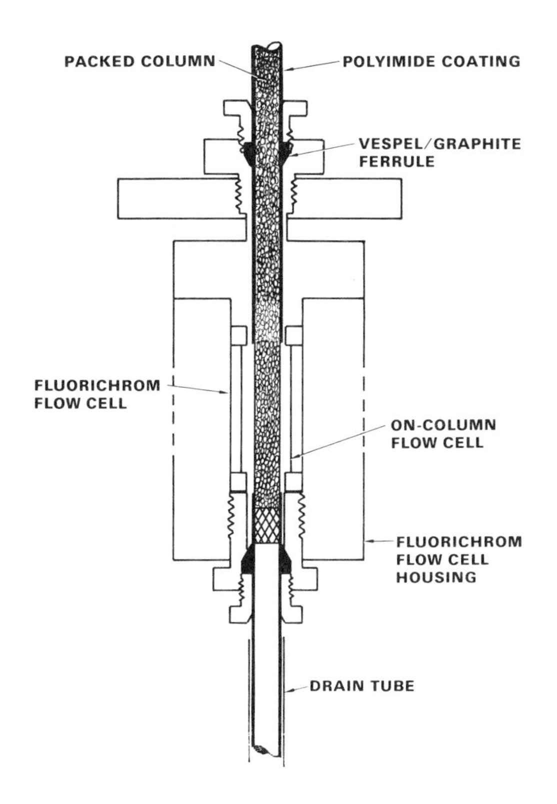

Figure 5. Schematic diagram for on-column fluorescence detection in conjunction with fused-silica narrow-bore packed column.

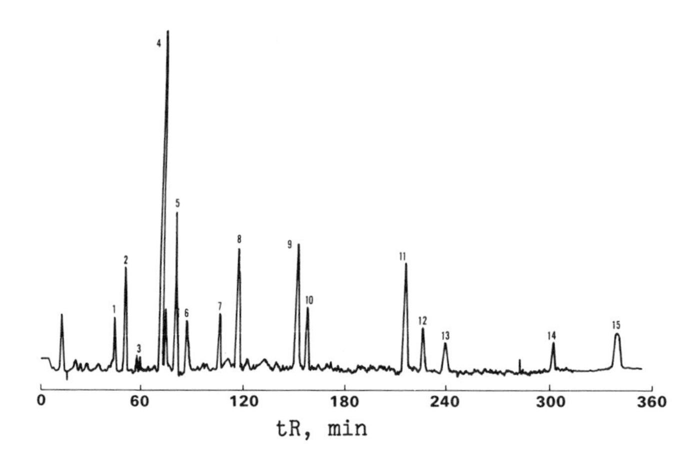

Figure 6. Gradient separations of PNA's on fused silica 66 cm x 250 μm, 5 μm Vydac 201 TP particles packed column. Mobile phases: (A) 65% acetonitrile, 35% water; (B) 80% acetonitrile, 20% dimethoxyethane. Sample components (ng injected): 1 naphthalene (20), 2 acenaphthylene (40), 3 acenaphthlene (20), 4 fluorene (4), 5 phenanthrene (2), 6 anthracene (2), 7 fluoranthene (4), 8 pyrene (2), 9 benz(a)anthracene (2), 10 chrysene (2), 11 benzo(b)fluoranthene (4), 12 benzo(k) fluoranthene (2), 13 benzo(a)pyrene (2, MDQ better than 50 pg), 14 dibenzo(a,h)anthracene (4), 15 benzo(ghi) perylene (4), 16 indeno (1,2,3-cd)-pyrene (2). (Reproduced with permission from Ref. 39, copyright Dr. Alfred Heuthig Publishers.)

the ion source pressure could be maintained constant over the entire measurement period. A flow rate below 15 μℓ/min allowed the source pressure to be kept constant to within less than 1% throughout a complete measuring cycle of 30 min. The interface was reported to work reliably over a period of 4 months, if high flow rates (i.e., >15 μℓ/min) are avoided. The development of narrow-bore packed columns (≤0.5 mm I.D.) which are operated at a low mobile phase flow rate (≤20 μℓ/min) is important in order to realize direct LC/MS interfacing.

Direct interfacing of microcapillary packed column LC to a GC flame-based detector (flame photometric detector and thermionic detector) was demonstrated by McGuffin and Novotny.[34,35] The flame-based detector was reported to accept in excess of 20 μℓ/min of 10-25% aqueous methanol without extinction of the flame. Optimum response was obtained at flow rates below 5 μℓ/min. Compatible solvent systems were aqueous methanol (up to 50%), acetone and ethanol (up to 40%). The minimum detectable quantity (at 5 times noise) measured for the FPD was 2 pg P. The dual-flame TSD can also be directly interfaced with microcapillary packed columns. The TSD was reported to be compatible with 75 to 100% aqueous methanol. The utilization of microbore column LC-TSD for the analyis of nitrogen, phosphorous, and halogen containing compounds is particularly important in studies of biomolecules, and drugs and their metabolites in physiological fluids.

Conclusion

Narrow-bore packed column HPLC offers advantages of both solvent saving and the separation power of columns over 100,000 theoretical plates. New spectroscopic detectors based on laser technology could improve the detection of narrow-bore packed column eluents. Most importantly, the potential of direct interfacing with mass spectrometer and flame-based detectors provide many additional possibilities in detection and applications. The practice of narrow-bore packed column HPLC requires substantial effort in the development of compatible hydraulic pumping systems. A state-of-the-art micro-HPLC pump must provide constant flow rates of ≥1 μℓ/min. Gradient elution in narrow-bore packed column HPLC requires a significant reduction in the gradient delay time of current systems and improvement in gradient linearity and reproducibility for the whole gradient range (0 to 100%). A microvolume high pressure (up to a few thousand bar) direct injection valve is desirable for small sample introduction to the narrow-bore packed columns. The on-column solute focussing technique should be applied in large sample injections. Narrow-bore packed column HPLC today can be practiced with some modifications of the conventional LC system. The utility of the narrow-bore packed column HPLC, however, now must be demonstrated in the applications of "real-life" samples.

Literature Cited

(1) M. Golay in "Gas Chromatography," 1978, Butterworth, London, 1959 (D.H. Desty ed.).

(2) K. Hibi, D. Ishii, I. Fujishima, T. Takeuchi and T. Nakanishi, J. High Resolu. Chromatogr. & Chromatogr. Commun., 1 (1978) 21.

(3) T. Tsuda, K. Hibi, T. Nakanishi, T. Takeuchi and D. Ishii, J. Chromatogr., 158 (1978) 227.

(4) T. Tsuda and G. Nakagawa, J. Chromatogr., 199 (1980) 249.

(5) F.J. Yang, J. High Resolu. Chromatogr. & Chromatogr. Commun., 3 (1980) 589.

(6) F.J. Yang, J. Chromatogr. Sci., 20 (1982) 241.

(7) R. Tijssen, J.P.A. Bleumer, A.L.C. Smit and M.E. Van Kreveld, J. Chromatogr., 218 (1981) 137.

(8) V. Pretorius and T.W. Smuts, Anal. Chem., 38 (1966) 274.

(9) G. Guiochon, Anal. Chem., 53 (1981) 1318.

(10) I. Halasz, J. Chromatogr., 173 (1979) 229.

(11) J.H. Knox and M.T. Gilbert, J. Chromatogr., 186 (1979) 405.

(12) F.J. Yang, J. Chromatogr., 236 (1982) 265.

(13) F.J. Yang, J. High Resolu. Chromatogr. & Chromatogr. Commun., 4 (1981) 83.

(14) T. Tsuda and M. Novotny, Anal. Chem., 50 (1978) 271.

(15) Y. Hirata, M. Novotny, T. Tsuda and D. Ishii, Anal. Chem., 51 (1979) 1807.

(16) R.P.W. and P. Kucera, J. Chromatogr., 185 (1979) 27.

(17) R.P.W. Scott, P. Kucera and M. Munroe, J. Chromatogr., 186 (1979) 475.

(18) P. Kucera and G. Manius, J. Chromatogr., 216 (1981) 9.

(19) C.E. Reese and R.P.W. Scott, J. Chromatogr. Sci., 18 (1980) 479.

(20) P. Kucera, J. Chromatogr., 198 (1980) 93.

(21) R.P.W. Scott, J. Chromatogr. Sci., 18 (1980) 49.

(22) R.P.W. Scott and P. Kucera, J. Chromatogr., 169 (1979) 51.

(23) R.A. Simpson, Walnut Creek Division, Varian Associates. Personal Commun.

(24) F.J. Yang, "Narrow-Bore Microparticle-Packed-Column HPLC," J. Chromatogr. Sci., in press.

(25) J.D. Hennion, Anal. Chem., 50 (1978) 1687.

(26) J.D. Hennion and G.A. Maylin, Biomed. Mass Spectrom., 7 (1980) 115.

(27) J.D. Hennion and T. Wachs, Anal. Chem., 53 (1981) 1963.

(28) J.D. Hennion, J. Chromatogr. Sci., 19 (1981) 57.

(29) K.H. Schafer and K. Levsen, J. Chromatogr., 206 (1981) 245.

(30) T. Takeuchi, Y. Hirata and Y. Okumura, Anal. Chem., 50 (1978) 659.

(31) Y. Hirata, T. Takeuchi, S. Tsuge and Y. Yoshida, Org. Mass Spectrom., 14 (1979) 126.

(32) S. Tsuge, Y. Hirata and T. Takeuchi, Anal. Chem., 51 (1979) 166.

(33) T. Takeuchi, D. Ishii, A. Saito and T. Ohki, J. High Resolu. Chromatogr. & Chromatogr. Commun., 5 (1982) 91.
(34) V.L. McGuffin and M. Novotny, Anal. Chem., 53 (1981) 946.
(35) V.L. McGuffin and M. Novotny, J. Chromatogr., 218 (1981) 179.
(36) T. Takeuchi and D. Ishii, J. High Resolu. Chromatogr. & Chromatogr. Commun., 4 (1981) 469.
(37) D. Ishii, K. Asai, K. Hibi, T. Jonokuchi and M. Nagaya, J. Chromatogr., 144 (1977) 157.
(38) T. Takeuchi and D. Ishii, J. Chromatogr., 253 (1982) 41.
(39) Sj. van der Wal and F.J. Yang, J. High Resolu. Chromatogr. & Chromatogr. Commun., 6 (1983) 216.
(40) S. Folestad, L. Johnson and B. Josefsson, Anal. Chem., 54 (1982) 925.
(41) R.A. Leach and J.M. Harris, J. Chromatogr., 218 (1981) 15.
(42) M.J. Pelletler, H.R. Thorshelm and J.M. Harris, Anal. Chem., 54 (1982) 239.
(43) S.D. Woodruff and E. Yeung, Anal. Chem., 54 (1982) 1174.
(44) M.J. Sepaniak and E.S. Yeung, J. Chromatogr., 211 (1981) 95.
(45) K. Jinno, C. Fujimoto and Y. Hirata, App. Spectrosc. 36 (1982) 67.
(46) K. Jinno, C. Fujimoto and D. Ishii, J. Chromatogr., 239 (1982) 625.
(47) T. Takeuchi and D. Ishii, J. Chromatogr., 213 (1981) 25.
(48) T. Takeuchi, Y. Watanabe, K. Matsuoka and D. Ishii, J. Chromatogr., 216, (1981) 153.
(49) T. Takeuchi and D. Ishii, J. Chromatogr., 238 (1982) 409.

RECEIVED November 3, 1983

9

Application of Micro High Performance Liquid Chromatography to the Separation of Complex Mixtures

DAIDO ISHII and TOYOHIDE TAKEUCHI

Department of Applied Chemistry, Faculty of Engineering, Nagoya University, Nagoya, 464 Japan

High-performance micro packed fused-silica and open-tubular glass capillary columns were prepared and applied to separations of complex mixtures. Solvent-gradient elution was quite useful for the separation of solutes with wide polarity. Instruments and some applications are described.

Nearly one decade has passed since the introduction of micro-scale high-performance liquid chromatography(HPLC)(1,2). Some advantages of micro HPLC arising from miniaturization as well as its efficiency have attracted a lot of chromatographers' attentions. Column preparation, injection and detection techniques suitable for micro HPLC have been developed and are improving.

The authors have recently employed fused-silica tubing as a column material and found that micro packed fused-silica columns possess high efficiencies(3-7), owing to the smooth and inert surface of fused-silica tubing. In previous work(4-6), 20-50 cm columns were connected in series in order to attain larger theoretical plate numbers. Theoretical plates around or in excess of 100,000 could be produced on 1.5-4 m micro packed fused-silica columns in the reversed-phase(5) and size-exclusion chromatography(4,6).

In gas chromatography(GC), open-tubular capillary columns can produce much more theoretical plate numbers than conventional packed columns. Open-tubular capillary columns have been examined as separation tools also in liquid chromatography(LC)(8-12). Coated columns, ca. 20 m X 10-30 μm i.d., produced theoretical plate numbers in excess of 100,000 but long-term stability of these columns was somewhat poor even if the mobile phase is saturated with the mobile phase. Accordingly chemically bonded or immobilized stationary phases should be preferred.

This paper describes the performance of micro packed fused-silica columns and chemically bonded octadecylsilane(ODS) open-tubular capillary columns.

0097–6156/84/0250–0109$06.00/0

Experimental

Column Preparation. Packed column : 10-50 cm micro packed fused-silica columns were manually prepared as described previously(3). 1-m Columns were prepared by the viscosity method using FAMILIC-300 (JASCO : Japan Spectroscopic, Tokyo, Japan) as the pump and a mixed solution of cyclohexanol, toluene and methanol as a slurry solution. In the former case, both ends of fused-silica tubing were inserted into narrow-bore(ca. 0.2 mm i.d.) PTFE tubing for small dead-volume connection. In the latter case, one end of fused-silica tubing was inserted in stainless-steel tubing, 0.5 mm i.d., 1/16" o.d. and 3-4 cm length. with an adhesive for high-pressure ferrule-type connection. In both cases, packings were prevented from leaking out with quartz wool.

Open-tubular column : Soda-lime glass capillaries were treated with a sodium hydroxide aqueous solution at 45-50 °C for 2 days and washed with methanol until the effluent became neutral. The capillaries were then dried under helium at 120 °C for a few hours and washed with methanolic 0.01 N hydrochloric acid, water and methanol. Finally, the capillaries were dried under helium at 120 °C for a few hours. Chemically and physically stable silica gel layer was formed on the glass surface with above pretreatments. An organic solution of octadecyltriethoxysilane was filled into the pretreated capillaries and the reaction was carried out at elevated temperature. After the reaction, columns were washed with organic solvents and the mobile phase. Both ends of columns were inserted into narrow-bore(ca. 0.2 mm i.d.) PTFE tubing for small dead-volume connection.

Apparatus. A Micro Feeder(Azumadenkikogyo, Tokyo, Japan) equipped with a gas-tight syringe(0.25 or 0.5 ml) was generally emplyed as a pump. This syringe-type pump endured about 70 kg/cm^2, being suitable for short packed columns and open-tubular capillary columns. FAMILIC-300 was also employed as a pump for operating a 1-m packed column at a high pressure. Small volume of sample solution could be injected with good reproducibility by a micro valve injector with 0.02 or 0.29 µl(JASCO). Solvent-gradient elution was performed with a home-made gradient equipment comprising a small-volume mixing vessel(0.064-0.11 ml) and a magnetic stirrer. Gradient profile was exponential, depending on the volume of the mixing vessel and flow-rate of the mobile phase(13). When necessary, the column temperature was kept constant with the use of a column oven. A UV spectrophotometer UVIDEC-100 (JASCO) equipped with a modified flow cell was generally emplyed as a detector. The detection volume of the flow cell was 0.04-0.1 µl.

Results and Discussion

Small consumption of expensive packing materials is one of advantages in micro HPLC. A micro column of around 10 cm X 0.25 mm i.d. requires only 10 mg or less of packing material. This makes it very convenient to examine the difference in the performance of packings originating from the difference in batches or manufaaturers. Figure 1 shows the performance of some packings commercially available. SC-01 was selected as the material in this work since this packing achieved lower HETP(height equivalent to a theoretical plate) values and lower dependence of HETP on flow-rates. Around 7,000 theoretical plates per 10 cm could be produced on an SC-01 column. In order to prepare the column rapidly acetonitrile was selected as a slurry solvent in this paper since it was difficult or time-consuming to employ the viscosity method for the manual packing. It may be possible to improve column efficiencies for other packing materials by selecting an appropriate slurry solution.

Solvent-gradient elution is the most popular mode of gradient elution in HPLC. However, at present, a gradient system for micro HPLC is not commercially available. The authors have recently developed a continuous gradient elution method for micro HPLC(13), in which the ratio of the flow-rate of the mobile phase to the volume of the mixing vessel determines gradient profile and an exponential gradient with an almost linear behavior in the beginning is performed. This gradient method was applied to the separation of complex mixtures using a 50-cm micro packed fused-silica column.

Figures 2 and 3 show gradient separations of epoxy resin oligomers, Epikote 1001 and 1004, respectively. Characterization of by-prducts as well as main peaks is of practical importance since the properties of epoxy resin are highly affected by them (14-16). Both ends of molecules occuring as main peaks are epoxide groups, while by-products have functional groups other than epoxide groups either as end group or as side chain pendant to the main chain. The structure of main peaks is shown in Scheme 1. Main peaks ranging from n=0 to n=13 appear in Figure 2, while main peaks ranging from n=0 to n=20 appear in Figure 3, owing to the difference in the degree of polymerization. Many peaks based on by-products are also resolved.

The pretreatment of a sample prior to injection is indispensable for the analysis of constituents existing in dilute solutions. In such a case, concentration of components using a micro precolumn method is effective, as reported previously(17,18). Organic constituents are commonly present at low levels in water. Thus, above micro precolumn method was useful for the analysis of these components. A precolumn employed in this work was composed of narrow-bore PTFE tubing(ca. 10 X 0.2 mm i.d.) packed with Develosil ODS-15/30(15-30 µm, Nomura Chemical, Seto-shi, Japan). After acetonitrile was passed through the precolumn for washing, an adequate amount of a sample solution was passed through the precolumn. The precolumn was then connected to the

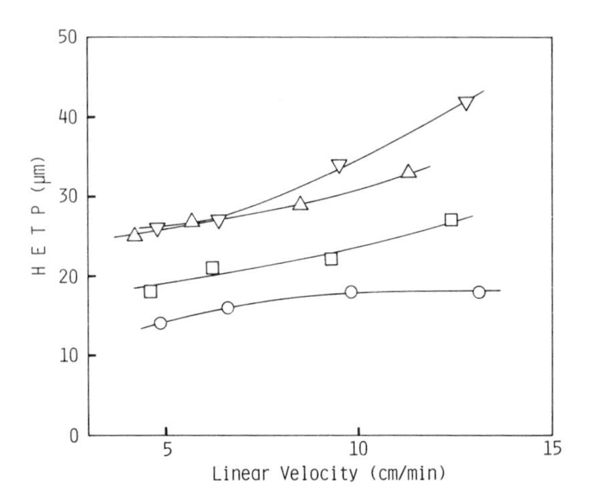

Figure 1. Performance of micro packed fused-silica columns. Pump:Micro Feeder. Column:10 cm X 0.22 mm i.d., = Lichrosorb RP-18(5 µm, Merck); =Nucleosil 5 C_{18}(5 µm, Machery-Nagel); =Partisil-5 ODS-3(5 µm, Whatman); = Silica ODS SC-01(5 µm, JASCO). Mobile phase:acetonitrile-water=70:30. Sample:pyrene. k'(column temperature): =7.5 (20 °C); =4.2(24 °C); =4.3(24 °C); =6.8(20 °C). Reproduced with the permission from Ref.7 Copyright 1983, Huethig.

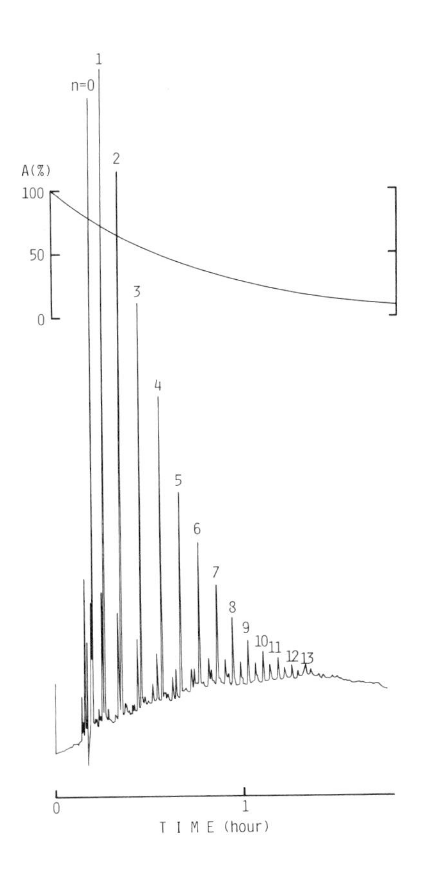

Figure 2. Gradient separation of Epikote 1001 on a micro packed fused-silica column. Pump:Micro Feeder. Column:SC-01, 50 cm X 0.22 mm i.d. Mobile phase:(A)acetonitrile-water=85:15; (B)acetonitrile-tetrahydrofuran=90:10, gradient profile as indicated. Flow-rate:1.4 µl/min. Sample: 0.16 µg of Epikote 1001. Wavelength:225 nm. Reproduced with the permission from Ref.7 Copyright 1983, Huethig.

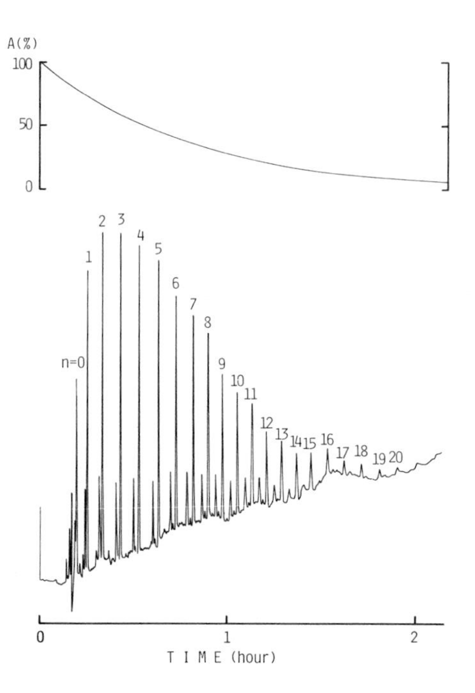

Figure 3. Gradient separation of Epikote 1004 on a micro packed fused-silica column. Operating conditions as in Figure 2 except the sample. Sample: 0.2 µg of Epikote 1004. Reproduced with the permission from Ref.7 Copyright 1983, Huethig.

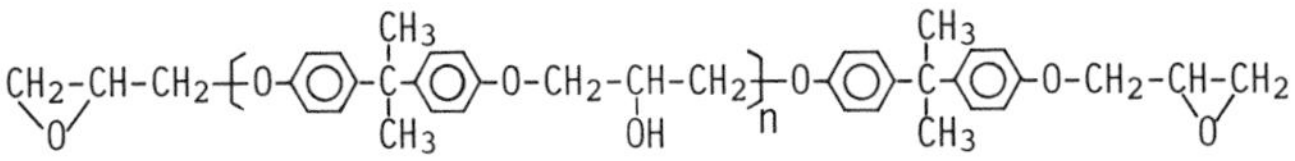

$CH_2(O)CH{-}CH_2{-}[O{-}C_6H_4{-}C(CH_3)_2{-}C_6H_4{-}O{-}CH_2{-}CH(OH){-}CH_2]_n{-}O{-}C_6H_4{-}C(CH_3)_2{-}C_6H_4{-}O{-}CH_2{-}CH(O)CH_2$

Scheme 1. Structure of Epikote oligomers.

separation column for the analysis. Figure 4 shows the gradient elution of constituents of distilled water in contact with powdered coal. A large number of peaks are seen due to high resolution prvided by the micro packed fused-silica column. The identification of these constituents is difficult, which may be solved by coupling high-resolution micro HPLC and mass spectrometry in the future.

Connection with PTFE tubing is unsuitable for the operation of high pressure. Thus, high-pressure ferrule-type connection of a fused-silica column was examined. Inserting fused-silica tubing in stainless-steel tubing with an adhesive was effective, which endured higher than 500 kg/cm^2. A prepared column was directly inserted in the body of the valve injector with a ferrule and a male nut, which enabled the operation at 300 kg/cm^2. Figure 5 demonstrates a separation of PCB(polychlorobiphenyl) mixtures on a 1-m column at optimum flow-rate in the constant-pressure mode. This technique will be useful for fast micro HPLC employing short columns as well as long columns.

In capillary GC, many kinds of immobilized stationary phases have been prepared, leading to increased column stability to temperature and organic solvents(19-23). However, a few papers on chemically bonded or immobilized statioanry phases have been reported in open-tubular capillary LC(9,24-28). The authors have examined polystyrene(26), ODS(27) and cation-exchange column(28) as chemically bonded or immobilized columns and found that the pretreatment described in the experimental section is necessary for preparing stable bonded phases. Figure 6 shows a gradient separation of epoxy resin oligomers on a 22 m X 31 μm i.d. ODS capillary column. The stationary phase was prepared in a stream of 10 %(V/V) toluene solution of octadecyltriethoxysilane for a few hours at 120 °C. Column temperature was kept at 52 °C. Resolution of main and by-products was obtained. Resolution of a prepared open-tubular ODS capillary column is somewhat poorer than that of micro packed fused-silica columns, as shown in Figures 2 and 6. Narrower-bore open-tubular ODS capillary columns will compete with the latter.

Conclusion

Micro packed fused-silica columns showed high resolution by resolving a lot of components of complex mixtures. Solutes with wide polarity were separated on high-resolution micro packed fused-silica columns with the help of gradient elution. Chemically bonded open-tubular capillary columns with 31 μm i.d. could resolve epoxy resin oligomers although their performance was somewhat poorer than that of packed columns.

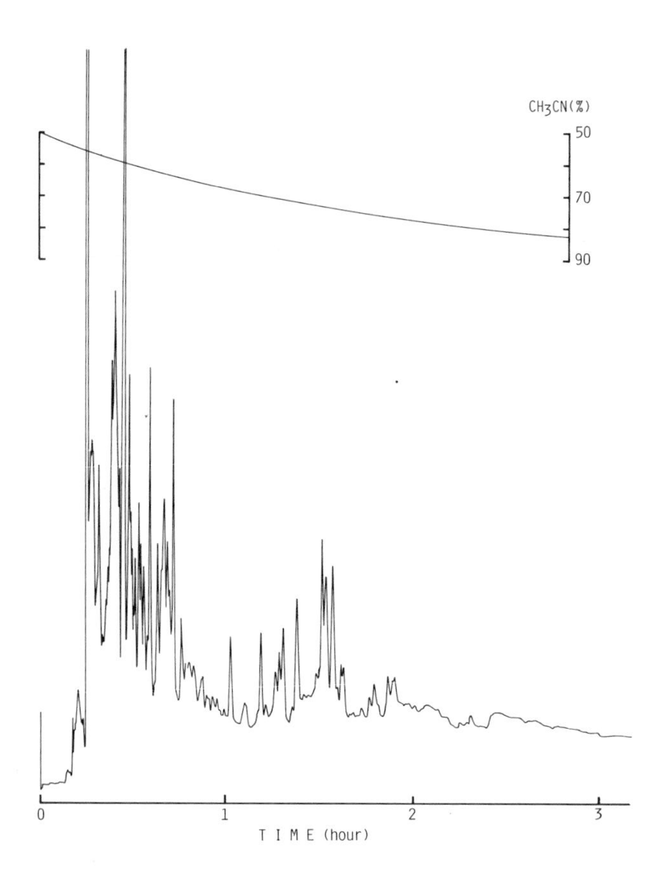

Figure 4. Analysis of water extracts from coal. Pump:Micro Feeder. Column:SC-01, 50 cm X 0.22 mm i.d. Mobile phase: acetonitrile-water, gradient profile as indicated. Flow-rate:1.04 µl/min. Sample:2.5 ml of distilled water in contact with powdered coal. Precolumn:Develosil ODS-15/30, 10 X 0.2 mm i.d. Wavelength of UV detection:225 nm. Reproduced with the permission from Ref.7 Copyright 1983, Huethig.

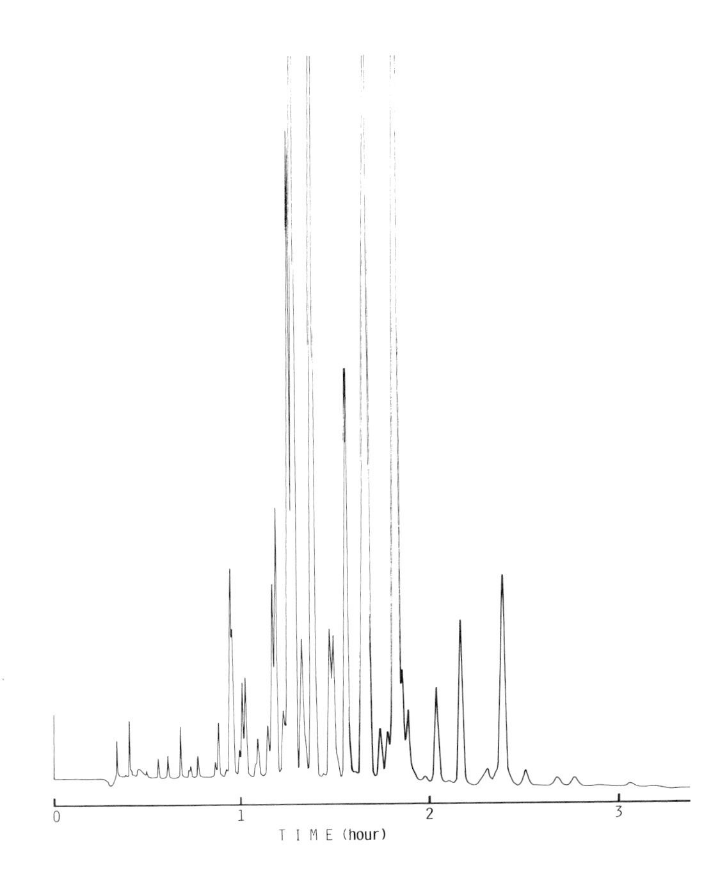

Figure 5. Separation of PCB mixtures on a 1-m micro packed fused-silica column. Pump:FAMILIC-300. Column:SC-01, 1 m X 0.34 mm i.d. Mobile phase:acetonitrile-water=85:15. Inlet pressure:80 kg/cm^2(2.8 µl/min). Sample:0.29 µl of 1 % PCB-48(chlorine content 48 %). Wavelength of UV detection:254 nm.

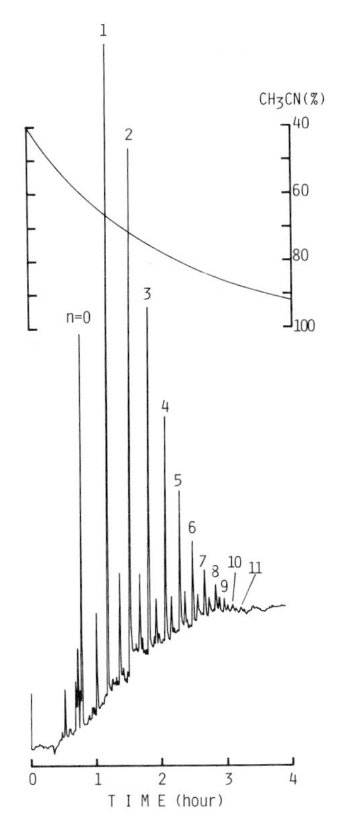

Figure 6. Gradient separation of Epikote 1001 on an open-tubular glass capillary column. Pump: Micro Feeder. Column: ODS. Mobile phase: acetonitrile-water, gradient profile as indicated. Flow rate: 0.52 μl/min. Sample: 0.16 μg of Epikote 1001. Wavelength of UV detection: 225 nm. Reproduced with permission from Ref. 7, Copyright 1983, Huethig.

Literature Cited

1. Scott, R.P.W.; Kucera, P.J. Chromatogr. 1976, 125, 251.
2. Ishii, D.; Asai, K.; Hibi, K.; Jonokuchi, T.; Nagaya, M. J. Chromatogr. 1977, 144, 157.
3. Takeuchi, T.; Ishii, D. J. Chromatogr. 1981, 213, 25.
4. Takeuchi, T.; Ishii, D. J. Chromatogr. 1982, 238, 409.
5. Ishii, D.; Takeuchi, T. J. Chromatogr. 1983, 255, 349.
6. Takeuchi, T.; Ishii, D.; Mori, S. J. Chromatogr. 1983, 257, 327.
7. Takeuchi, T.; Ishii, D. J. High Resolut. Chromatogr. Chromatogr. Commun. 1983, 6, 310.
8. Hibi, K.; Tsuda, T.; Takeuchi, T. Nakanishi, T.; Ishii, D. J. Chromatogr. 1979, 175, 105.
9. Ishii, D.; Takeuchi, T. J. Chromatogr. Sci. 1980, 18, 462.
10. Tsuda, T.; Tsuboi, K.; Nakagawa, G. J. Chromatogr.1981, 214, 283.
11. Tijssen, R.; Bleumer, J.P.A.; Smit, A.L.C.; Krevelt, M.E. van. J. Chromatogr. 1981, 218, 137.
12. Crejčí, M.; Tesařík, K.; Rusek, M.; Pajurek, J. J. Chromatogr. 1981, 218, 167.
13. Takeuchi, T.; Ishii, D. J. Chromatogr. 1982, 253, 41.
14. Dark, W.A.; Conard, E.C.; Crossman, L.W.Jr. J. Chromatogr. 1974, 91, 247.
15. van der Maeden, F.P.B.; Biemond, M.E.F.; Janssen, P.C.G.M. J. Chromatogr. 1978, 149, 539.
16. Shiono, S.; Karino, I.; Ishimura, A.; Enomoto, J. J. Chromatogr. 1980, 193, 243.
17. Ishii, D.; Hibi, K.; Asai, K.; Nagaya, M.; Mochizuki, K.; Mochida, Y. J. Chromatogr. 1978, 156, 173.
18. Takeuchi, T.; Ishii, D. J. Chromatogr. 1981, 218, 199.
19. Blomberg, L.; Wännman, T. J. Chromatogr. 1979, 168, 81.
20. Grob, K.; Grob, G. J. Chromatogr. 1981, 213, 211.
21. Blomberg, L.; Buijten, J.; Markides, K.; Wännman, T. J. Chromatogr. 1982, 239, 51.
22. Lipsky, S.R.; McMurray, W.J. J. Chromatogr. 1982, 239, 61.
23. Peaden, P.A.; Wright, B.W.; Lee, M.L. Chromatographia 1982, 15, 335.
24. Tsuda, T.; Hibi, K.; Nakanishi, T.; Takeuchi, T.; Ishii, D. J. Chromatogr. 1978, 158, 227.
25. Yang, F.J. J. High Resolut. Chromatogr. Chromatogr. Commun. 1980, 3, 589.
26. Takeuchi, T.; Matsuoka, K.; Watanabe, Y.; Ishii, D. J. Chromatogr. 1980, 192, 127.
27. Ishii, D.; Takeuchi, T. Proceeding of IVth International Symposium on capillary chromatography 1981, p.11.
28. Ishii, D.; Takeuchi, T. J. Chromatogr. 1981, 218, 189.

RECEIVED October 13, 1983

10

High Resolution Supercritical Fluid Chromatography

W. P. JACKSON, B. E. RICHTER, J. C. FJELDSTED, R. C. KONG, and M. L. LEE

Department of Chemistry, Brigham Young University, Provo, UT 84602

Supercritical fluid chromatography (SFC) has been studied for a number of years, but is presently undergoing more rapid growth. There are several features which make SFC desirable. The properties of a supercritical fluid are intermediate to those of gases and liquids. Diffusion coefficients are higher in supercritical fluids than in liquids, thus leading to higher efficiency and increased speeds of analysis. Furthermore, the greater density of supercritical fluids over gases gives solvating properties to these fluids, and high molecular weight and thermally labile compounds can be analyzed.

Capillary columns are desirable in SFC because the pressure drop across the column can be controlled to a minimum and high efficiencies can be obtained. Practical considerations of capillary SFC, including column technology and instrumentation, are reviewed in this paper.

Although supercritical fluid chromatography (SFC) has been around since 1962, it has found little interest and use as an analytical technique (until recently) because of practical limitations in the areas of column technology (efficiency and stability), and high pressure instrumentation (most importantly, pressure regulation and detector design). The potential advantages of SFC when compared to gas and liquid chromatography stem from (a) higher solute diffusivities in supercritical fluids than in liquids, (b) lower viscosities than liquids, and (c) excellent solvating properties. These properties should lead to higher chromatographic efficiencies than are obtainable in liquid chromatography (although not as high as in gas chromatography), faster analysis times than are

0097-6156/84/0250-0121$06.00/0

obtainable in liquid chromatography, and the possibility of separating high-molecular-mass molecules (far beyond the volatility range of gas chromatography) at relatively low temperatures.

The first report of the use of SFC was in the separation of metal phorphyrins (1). In this study, and in all studies up until 1981, packed columns were utilized. In 1981, the first use of open tubular (capillary) columns in SFC was reported (2). Uncoated capillary columns were used prior to this time to measure binary diffusion coefficients (3-5), but not for chromatography. The main benefit in using capillary columns is the same as for gas chromatography; because of the openness of the column, long column lengths can be used and large numbers of theoretical plates obtained. Recently, the methods and principles of SFC, including the use of capillary columns was reviewed (6). Since the first report of capillary SFC, a number of significant advances have been made, and are discussed in the following sections.

Column Technology

As is the case for both gas and liquid chromatography, the analytical column is the heart of the SFC technique. The quality of the separation can be no better than the quality of the column. This is normally measured in terms of efficiency, deactivation, and stationary phase film stability. Column technology for capillary SFC more closely resembles that for capillary GC than for HPLC; the column wall should be deactivated and coated with a thin film of stationary phase, after which the stationary phase is rendered nonextractable by free-radical crosslinking.

Coating of capillary columns with stationary phases for SFC can be more difficult than for conventional capillary GC. This is because smaller column diameters are required. While extremely large numbers of theoretical plates are possible with larger diameters (7,8), calculations from chromatographic theory of the internal diameters and column lengths necessary to achieve relatively high efficiencies in reasonable analysis times indicate that column diameters of 50 to 100 µm i.d. are necessary for high-resolution SFC (8). For example, more than 10^5 effective theoretical plates (k'=5) are possible in approximately two hours on 30-m long columns of 50 µm i.d. (carrier velocity = 2.4 cm s^{-1}). In order to realize the goal of high resolution in SFC, technology must be available to efficiently coat narrow-bore columns.

Capillary columns are normally coated statically at temperatures below the atmospheric boiling point of the solvent that was used to prepare the coating solution. At these temperatures, coating of small diameter capillaries is difficult, requiring extremely long time periods (several days). These long coating times increase the possibility of variation in coating conditions (e.g., temperature of the coating bath) which often result in nonuniform film deposition and even plugging of the column. Recent results (9) indicate that much higher coating temperatures, above the boiling point of the solvent, can be used to prepare efficient capillary columns in much shorter times (as much as 80% reduction in time). This discovery has greatly facilitated the preparation of small-diameter columns for SFC. Efficient 50 µm i.d. columns as long as 60 meters (0.25 µm film thickness) have been produced.

As mentioned earlier, the stationary phase must also be rendered nonextractable under supercritical fluid conditions. This has been accomplished by in situ free-radical crosslinking. Excellent results have been obtained using azo-t-butane as free-radical generator (10). Azo-t-butane is superior to peroxides because its decomposition products are nonpolar (nitrogen gas and t-butane), and it does not promote oxidation of susceptible stationary phases with consequent activity and deterioration (11). Other azo compounds are currently under investigation as well. Using azo-t-butane, both apolar and polar nonextractable phases have been produced for small-bore columns.

Instrumentation

SFC instrumentation is, in many respects, a hybrid of GC and HPLC components. Figure 1 shows a schematic diagram of the capillary SFC system. Pressure is maintained using an HPLC syringe pump which has been modified for pressure control (12). The temperature of the column is controlled above and near the critical temperature (usually at a reduced temperature of 1.03) using a GC oven. The temperature is held constant and the pressure is increased to effect programmed elution. Sample introduction, density programming, and detection are discussed in the following sections.

Sample Introduction. The small internal diameters of SFC capillary columns place stringent requirements on sample introduction in order to avoid band-broadening due to too large of an injector volume. These requirements have been met using a 0.2-µL internal volume high pressure valve operated in a splitting mode (12). Split ratios of 3:1 to 5:1 are usual. Recently, valves designed specifically for use with capillaries have become available (Valco Instruments Co., Houston, Texas). These valves offer internal volumes as low as 60-nL and show promise for use in capillary SFC.

Density Control and Programming. Since the pressure drop across the column length decreases the selectivity and, hence, resolution of mixture components (6,8,13), it is very important to minimize the pressure drop across the column using a restrictor at the column outlet. Typically, the desired restriction is achieved

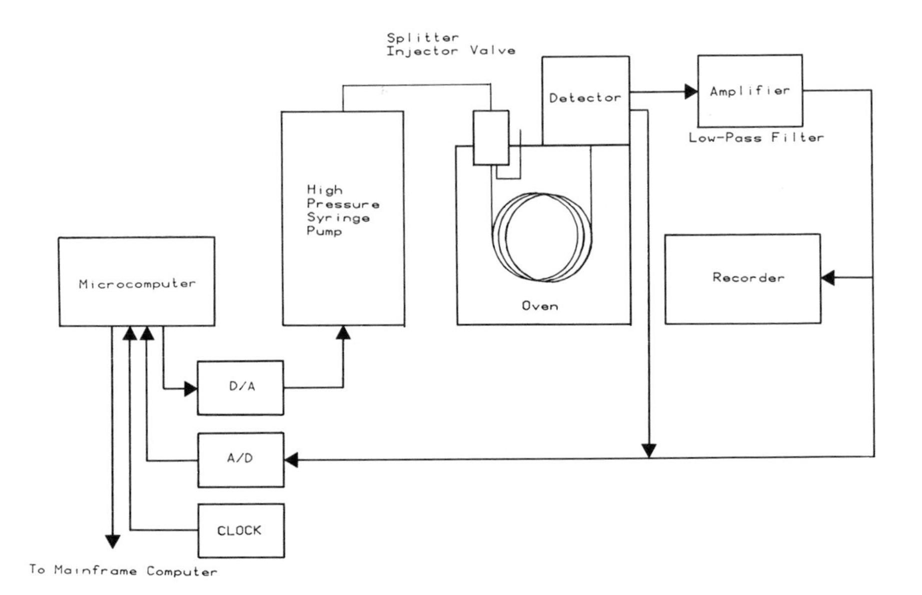

Figure 1. Schematic diagram of capillary supercritical fluid chromatographic instrumentation.

using a short length of 10 μm i.d. fused silica tubing (Scientific Glass Engineering). In order to maintain the desired mobile phase density and to ensure reproducibility, micro-computers have been interfaced to the syringe pumps used to deliver the mobile phase (14). A simple equivalent to temperature programming in gas chromatography and solvent programming in liquid chromatography is available in SFC by pressure programming. As the pressure is increased, the fluid density approaches that of a liquid, and the mobile phase solvating power increases. This allows a large molecular mass range to be analyzed in a single chromatographic run.

Various density and pressure programming schemes have been investigated (14). Figure 2(A) shows a separation of a polystyrene oligomer mixture with a nominal molecular mass of 2000 daltons. For this separation, the pressure was increased as a linear function of time. The solvating capacity of the supercritical fluid is not directly related to pressure, and density programming is more appropriate than pressure programming (14,15). Figure 2(B) shows a separation of the same 2000 MW polystyrene oligomers using an asymptotic density program (14). Note the more even spacing of the polystyrene oligomers using this density program. While not indicative of the efficiencies that can be achieved in capillary SFC, these chromatograms (Figure 2) illustrate the usefulness of density programming.

Detection. An important aspect of capillary SFC is the availability of many different detection systems. Proper choice of the SFC mobile phase makes possible the use of nearly all of the detectors commonly used in both HPLC and GC.

UV-absorption detectors are commonly used in HPLC and can be used in SFC. On-column detection in capillary SFC using UV-absorption has been reported (14). The polystyrene separations referred to in the previous section were accomplished using a UV-absorption detector. The sensitivity of the system is approximately 30 pg s^{-1}. This system has also been applied to the detection of high-molecular-mass polycyclic aromatic compounds (PAC). Figure 3 shows a capillary chromatogram of a carbon black extract with UV absorption monitored at 214 nm. While GC can be used to separate PAC of up to 7 rings (∿ 300 daltons), supercritical n-pentane has extended the analysis to compounds containing more than 10 rings (400+ daltons). The carbon black extract contained compounds of molecular weight extending only to approximately 400 daltons. It should be possible to chromatograph much larger compounds as evidenced by the polystyrene analysis.

Fluorescence detection, another common HPLC detection method, has also been used in capillary SFC (2,12,16). Fluorescence detection offers several advantages. First, sensitivity is much greater than in UV-absorption, on the order of 500 fg s^{-1} with a fiber optics arrangement (16). The very low background noise in

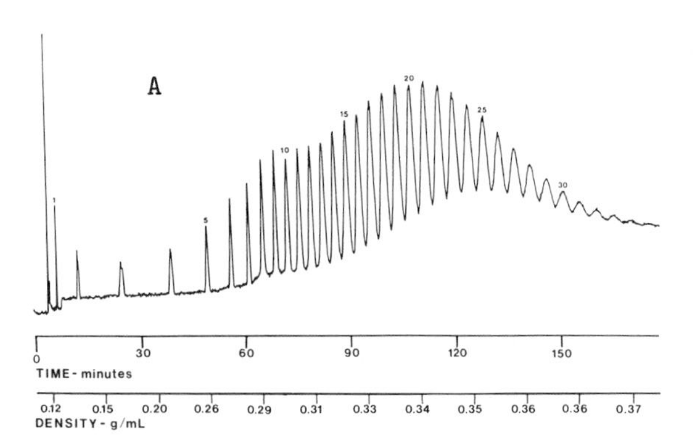

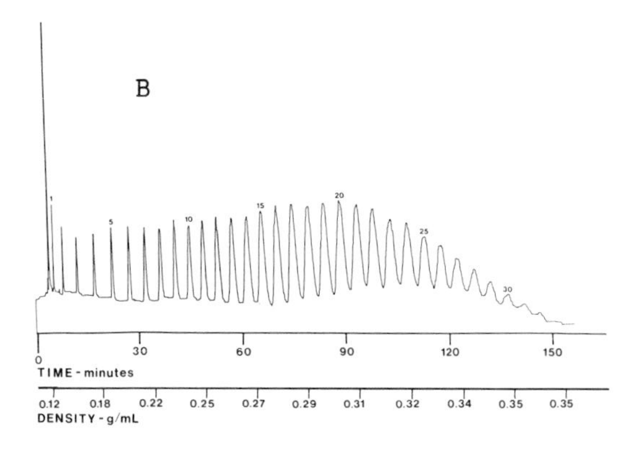

Figure 2. SFC chromatograms of a 2000 $\overline{MW}$ polystyrene oligomer mixture. Conditions: n-pentane mobile phase at 210°C, 10 m x 100 μm i.d. fused silica capillary column coated with a 0.25-μm film of crosslinked 50% phenyl methylphenylpolysiloxane, UV-absorption at 205 nm, (A) linear pressure program from 34.6 atm (0.12 g/mL) to 60 atm at 0.15 atm/min after a 5-min isobaric period (B) asymptotic density program from 0.12 g/mL (34.6 atm) to 0.35 g/mL according to the equation $\rho = 0.48 - \frac{28.8}{t+80}$ after a 5-min isoconfertic (constant density) period.

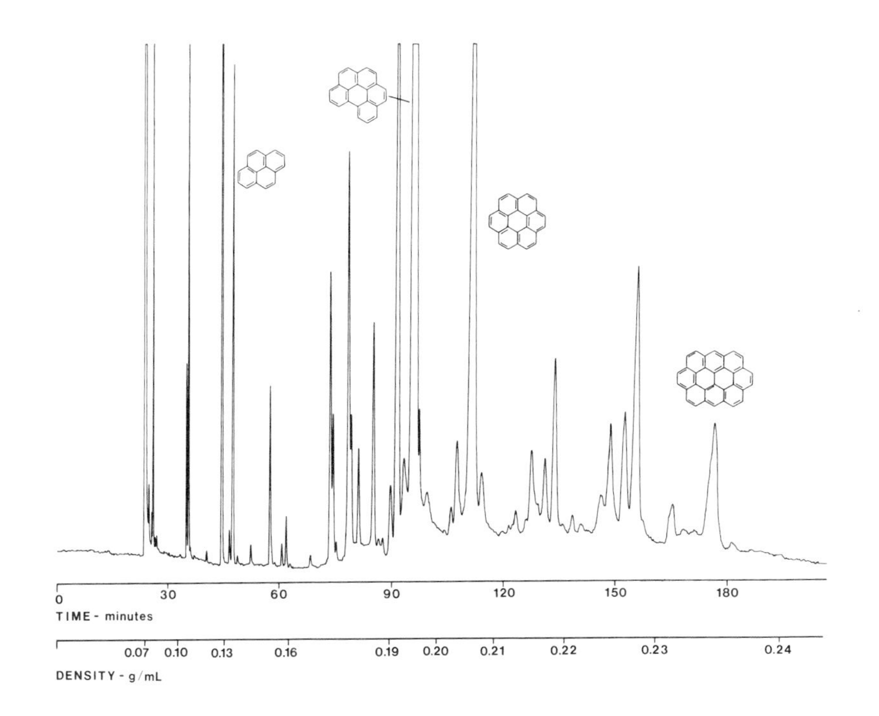

Figure 3. SFC chromatogram of a carbon black extract. Conditions: n-pentane mobile phase at 210°C, 32 m x 75 μm i.d. fused silica capillary column coated with a 0.25-μm film of crosslinked 50% phenyl methylphenylpolysiloxane, UV-absorption at 214 nm, asymptotic density program from 0.07 g/mL (26.7 atm) to 0.24 g/mL according to the equation $\rho = 0.30 - \frac{13.8}{t+60}$ after a 24-min isoconfertic period.

this detection system has also made possible microcomputer-controlled fluorescence scanning. Figure 4 shows the total fluorescence chromatogram (analogous to a total ion plot in GC-MS) of a carbon black extract. The fluorescence scanning and computer-controlled data acquisition make possible reconstruction of total fluorescence chromatograms as well as selected wavelength chromatograms. In addition, fluorescence scanning makes possible the collection of on-the-fly spectra of eluting compounds. Fluorescence spectra can thus aid in identification of resolved compounds.

The majority of work done with HPLC detectors (i.e., UV-absorption and fluorescence) has been accomplished using n-pentane as the mobile phase. Several other mobile phases are currently under investigation (isobutane, n-butane, benzene, methanol, isopropanol, ammonia, carbon dioxide, and nitrous oxide).

Carbon dioxide and nitrous oxide have made possible the use of conventional GC flame detectors (17). While specific design modifications were essential for packed column SFC (18), the detector configuration used for capillary SFC was essentially the same as in GC (except for the added short 10 μm i.d. fused silica restrictor at the column end for pressure reduction before the flame), as were the resultant sensitivity and response characteristics. The availability of FID in capillary SFC provides a sensitive, universal detector, the lack of which is a major drawback in HPLC. Figure 5(A) shows a chromatogram of the aliphatic fraction of a solvent refined coal product. This separation was accomplished with supercritical CO_2 at 40°C and conventional FID detection. For comparison, a capillary gas chromatogram of the same sample with temperature programming to 265°C is shown in Figure 5(B). It is apparent that the efficiencies of SFC are approaching those of GC. Plate heights of 0.30 mm have been obtained in capillary SFC on 50 μm i.d. columns with supercritical CO_2 at 40°C. This compares favorably with the 0.20-0.25 mm plate heights commonly obtained in capillary GC using 0.25-0.30 mm i.d. columns. This comparison is even more remarkable when note is taken of the fact that the SFC separations were obtained at 40°C.

The low critical temperatures for some mobile phases used in SFC separations make this analytical technique especially desirable for analysis of thermally labile compounds (17). It is expected that capillary SFC will replace many GC and HPLC analyses in which compounds must be derivatized prior to analysis. For example, various thermally labile azo compounds have been separated using supercritical CO_2 at 40°C (17). Also, free cholesterol has been successfully eluted using supercritical N_2O at 46°C (17).

In addition to universal detection with FID, element-specific detection is also possible. Just as selective detection has become invaluable in GC, it should become equally as powerful in SFC. Nitrogen-specific analyses using an NPD have been reported (17). The detector characteristics (i.e., sensitivity, linearity,

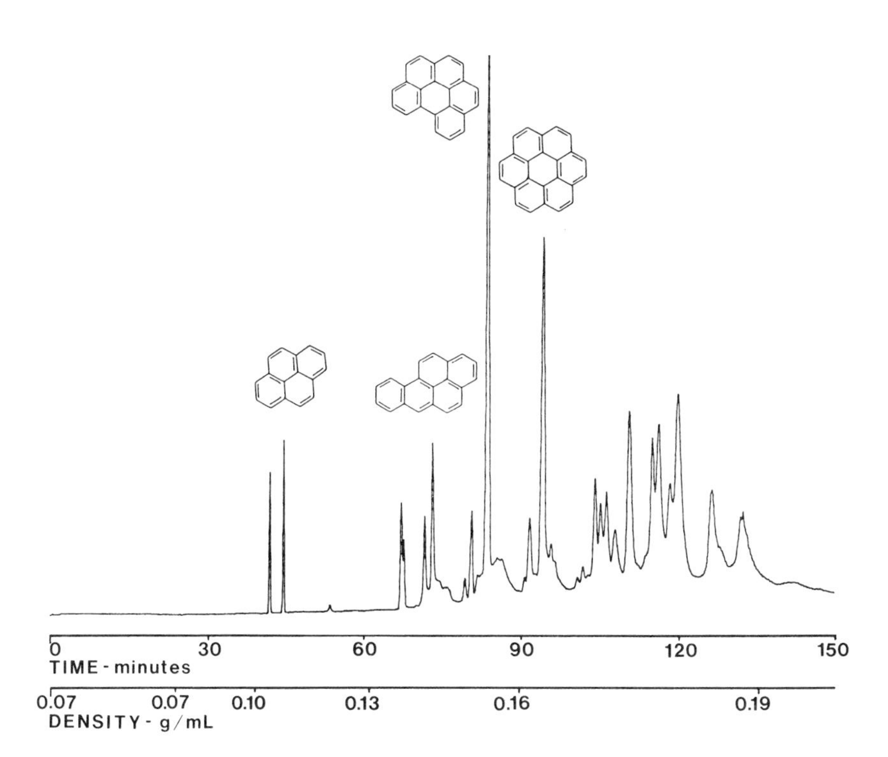

Figure 4. Reconstructed total fluorescence SFC chromatogram of a carbon black extract. Conditions: n-pentane mobile phase at 210°C, 32 m x 75 µm i.d. fused silica capillary column coated with a 0.25-µm film of crosslinked 50% phenyl methylphenylpolysiloxane, fluorescence excitation scanned from 300 to 420 nm (10 nm slit), fluorescent emission monitored at 460 nm (20 nm slit), asymptotic density program from 0.07 g/mL (26.7 atm) to 0.19 g/mL according to the equation $\rho = 0.30 - \frac{23}{t+100}$ after a 25-min isoconfertic period.

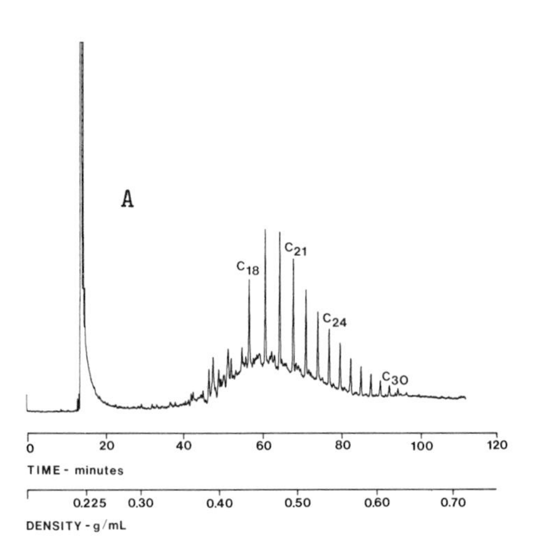

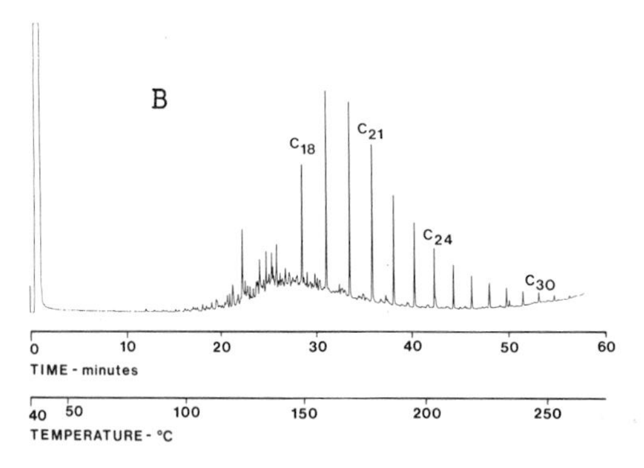

Figure 5. Capillary chromatograms of an aliphatic fraction from a solvent refined coal liquid. Conditions: (A) SFC chromatogram, CO_2 mobile phase at 40°C, 34 m x 50 µm i.d. fused silica capillary column coated with a 0.25-µm film of crosslinked SE-54, FID detector, linear density program from 0.225 g/mL (73 atm) to 0.70 g/mL (112 atm) at 0.005 g/mL-min after a 15-min isoconfertic period, (B) GC chromatogram, H_2 carrier gas, 20 m x 300 µm i.d. fused silica capillary column coated with a 0.25-µm film of crosslinked SE-54, FID detector, temperature program from 40°C to 265°C at 2°C/min after a 4-min isothermal period.

selectivity, stability, etc.) are essentially identical to what is obtained in GC.

Another advantage of capillary SFC is the ease with which it can be coupled to a mass spectrometer (MS). HPLC and packed column SFC are hampered by large volumes of solvent (mobile phase) which prevent direct interface to the MS and necessitate elaborate measures to remove a large portion of the mobile phase. However, the small volumes (mass flow) common to capillary SFC make a direct interface, similar to GC-MS, possible. Results (19,20) have shown it to be straightforward to interface capillary SFC to a mass spectrometer. The end of the capillary with its restrictor was incorporated in the direct fluid injector probe leading to the ion source of a conventional chemical ionization mass spectrometer. Isobutane and *n*-pentane, which can serve as chemical ionization reagents themselves, have been used thus far as the principal mobile phases for SFC-MS. For the same reasons that GC-MS has proven to be one of the most powerful analytical tools, SFC-MS should extend the usefulness of chromatography-mass spectrometry systems to a wide range of compounds which presently cannot be analyzed using these techniques.

Summary

Capillary SFC is still in its infancy, but current results indicate its future potential. Although small diameter (< 100 μm) columns are essential, column technology is readily adaptable from capillary GC to produce efficient, nonextractable stationary phases. The high resolution obtainable, plus the capability of coupling capillary SFC to most of the conventional HPLC and GC detectors make this analytical technique a highly versatile method of analysis for a wide range of applications. The detectors that have been successfully used include UV-absorption, fluorescence, flame ionization, and nitrogen thermionic. In addition, it has been rather straightforward to interface capillary SFC to mass spectrometry. This powerful combination is not plagued with the usual problems of solvent removal as encountered in HPLC-MS. There is every reason to expect that SFC will eventually replace a large fraction of the analyses presently done by GC and HPLC. Capillary SFC has already begun to show its potential in many applications such as polymers, large coal molecules, and thermally labile compounds. Recent developments and improvements in SFC instrumentation, including the ready availability of micro-computer control make the technique even more desirable and accessible.

Most of the work to date has been done using relatively few mobile phases (*n*-pentane, carbon dioxide, and nitrous oxide). Future efforts are needed to explore the use of other more polar and/or mixed mobile phases. Furthermore, column surface deactivation has become more important as emphasis has focussed more on the analysis of polar and trace compounds. The greatly increasing complexities of sample compositions as analyses are extended to high-molecular-mass compounds necessitates the development of more sensitive and selective detection systems. It is anticipated that much future effort will be concentrated in these areas.

Since the initial discovery of SFC, this analytical technique has enjoyed rather slow growth. It is believed that the solution of a number of technological problems by the introduction of capillary columns in SFC will help to accelerate this growth. As additional problems are solved and new applications are demonstrated, there is every reason to expect that capillary SFC will eventually become widespread.

Acknowledgments

This work was supported by the Gas Research Institute (GRI), Contract No. 5081-260-0586, and the U.S. Department of Energy (DOE), Grant No. DE-FG22-81PC40809. Any opinions, findings, conclusions or recommendations expressed herein are those of the authors and do not necessarily reflect the views of DOE or GRI.

Literature Cited

1. E. Klesper, A.H. Corwin, and D.A. Turner, J. Org. Chem. 27, 700 (1962).
2. M. Novotny, S.R. Springston, P.A. Peaden, J.C. Fjeldsted, and M.L. Lee, Anal. Chem. 53, 407A (1981).
3. U. van Wasen, Dissertation, Ruhr-Universität, Bochum, 1978.
4. I. Swaid and G.M. Schneider, Ber. Bunsenges. Phys. Chem. 83, 969 (1979).
5. R. Feist and G.M. Schneider, Sep. Sci. Technol. 17, 261 (1982).
6. P.A. Peaden and M.L. Lee, J. Liq. Chromatogr. 5, 179 (1982).
7. S.R. Springston and M. Novotny, Chromatographia 14, 679 (1981).
8. P.A. Peaden and M.L. Lee, J. Chromatogr. 259, 1 (1983).
9. R.C. Kong and M.L. Lee, J. High Resoln. Chromatogr./ Chromatogr. Commun., 6, 319 (1983).
10. B.W. Wright, P.A. Peaden, M.L. Lee, and T. Stark, J. Chromatogr. 248, 17 (1982).
11. B.E. Richter, J.C. Kuei, J.I. Shelton, L.W. Castle, J.S. Bradshaw, and M.L. Lee, J. Chromatogr., in press.
12. P.A. Peaden, J.C. Fjeldsted, M.L. Lee, S.R. Springston, and M. Novotny, Anal. Chem. 54, 1090 (1982).
13. J.A. Graham and L.B. Rogers, J. Chromatogr. Sci. 18, 75 (1980).
14. J.C. Fjeldsted, W.P. Jackson, P.A. Peaden, and M.L. Lee, J. Chromatogr. Sci. 21, 222 (1983).
15. L.M. Bowman, Jr., M.N. Myers, and J.C. Giddings, Sep. Sci. Technol. 17, 271 (1982).

16. J.C. Fjeldsted, B.E. Richter, W.P. Jackson, and M.L. Lee, J. Chromatogr., in press.
17. J.C. Fjeldsted, R.C. Kong, and M.L. Lee, J. Chromatogr., in press.
18. O. Vitzthum, P. Hubert, and M. Barthels, U.S. Patent 3,827,859, Aug. 6, 1974.
19. R.D. Smith, W.D. Felix, J.C. Fjeldsted, and M.L. Lee, Anal. Chem. 54, 1883 (1982).
20. R.D. Smith, J.C. Fjeldsted, and M.L. Lee, J. Chromatogr. 237, 231 (1982).

RECEIVED September 28, 1983

11

Carbon Dioxide Based Supercritical Fluid Chromatography

Column Efficiencies and Mobile Phase Solvent Power

L. G. RANDALL

Hewlett-Packard, Avondale, PA 19311

Efficiencies attainable in supercritical fluid chromatography (SFC) are compared to those encountered in packed column high performance liquid chromatography (HPLC) and capillary column gas chromatography (GC) systems. Since SFC efficiencies are quite acceptable for high resolution chromatography, exploration of of the available solvent power range of supercritical fluid mobile phases is warranted. In particular, the application of the Snyder solvent classification scheme recently used for the optimization of LC separations is proposed as a framework to aid the supercritical fluid chromatographer in the selection of appropriate modifiers to alter the solvent power of carbon dioxide. Preliminary experiments to test the feasibility of this approach have shown that the chromatographic capacity factors and separation ratios can be greatly affected by not only the modifier identity but also the modifier concentration in the modifier/carbon dioxide solvent mixture.

One of the most compelling reasons to use supercritical fluids as solvents in extractions and chromatographic systems is that the solvent powers of these fluids approach those of typical liquids while the viscosities are gas-like and the diffusivities can approach values intermediate between typical gases and liquids. Moreover, the supercritical fluid solvent power is variable, being a nearly linear function of the fluid density, so a wide range of solvent powers is available for each supercritical fluid solvent. This means that a physical parameter, the density (or pressure), can be varied to change the solvent power. For those who work primarily with gas chromatography (GC) where the mobile phase is used only for zone movement, and temperature, not the mobile phase composition, is the parameter used to control analyte concentration in the mobile phase, the importance of a variable solvent power coupled with transport properties somewhat less than favorable for GC systems may not be readily apparent. However, it has been noted

0097-6156/84/0250-0135$09.50/0

that only about 10% of the two million known compounds are amenable to analysis by GC (1). High performance liquid chromatography (HPLC) is used to separate many mixtures that cannot be analyzed by GC because the mobile phase serves two purposes: solvation and zone movement. Because of the mobile phase interaction with the solute and the mild operating temperatures of HPLC, HPLC is suitable for analysis of involatile and/or thermally labile compounds (the "other 90%"). With a supercritical fluid being a hybrid of gases and liquids as we normally encounter them, it follows that supercritical fluid chromatography (SFC) is a chromatographic technique that is a combination of and complementary to GC and HPLC. When making the decision as to which of these three chromatographic techniques to use for a separation, the investigator must consider analyte volatility and thermolability, mobile phase solvent power (if any), the required chromatographic efficiency, and finally the speed of the analysis.

As the title implies, the main purpose of this work was to study mobile phase solvent power. This results from the fact that, even though a supercritical fluid has a variable solvent power, there is a maximum value of the solvent power--essentially that of the substance as a liquid. If that maximum solvent power is not high enough, a higher value can be achieved by using a mixture of the chosen supercritical fluid with a second, higher-solvent-strength component. The second component is often referred to as a "modifier."

Once the necessary mobile phase solvent power has been established (implying prior elimination of the GC choice), the chromatographic efficiency and speed of analysis should be considered in making the choice between capillary column SFC, packed column SFC, and conventional packed column HPLC. Before exploring the specific topic of mobile phase composition in carbon dioxide based chromatography, the question of whether the realm of ultrahigh resolution chromatography is accesible to SFC should be considered.

EFFICIENCIES IN SFC

It is well-known that extremely high efficiencies are possible in capillary column GC: e.g., 250,000 total plates for a 50 m column yielding 5,000 plates/meter (2,3). By comparison, in HPLC, classification of columns as to high resolution or ultrahigh resolution appears to be less common. A discussion about what parameters are important in characterizing column performance has been presented by Snyder and Kirkland (4). In particular, they outline the reduced parameter approach of Giddings (5), where the use of the reduced plate height ($h = HETP/d_p$) and the reduced mobile phase velocity ($\nu = \bar{u}d_p/D_{12}$) yields one general relationship. (The terms are defined as follows: h--reduced plate height, HETP--height equivalent to a theoretical plate, d_p--particle diameter, ν --reduced mobile phase linear velocity, $\bar{u}$--mobile phase average linear velocity and D_{12} --binary diffusivity.) A very important point is that the reduced plate height is a function of only the reduced velocity (ν) and not of packing particle diameter. Furthermore, the plate height coefficients for the contributions of the different band broadening processes to the column plate height have constant, minimum values that are also independent of particle diameter from column to column for "well-packed" columns. A graphical presentation yields an optimum reduced velocity of about 3 and a minimum reduced plate height of about 2 for all columns. If it can be assumed that particle diameters of 3 μm are presently state-of-the-art for LC systems, then the limiting minimum plate height (or height equivalent to a theoretical plate) is 6 μm and a 25 cm packed HPLC column would have 42,000 plates, corresponding to 168,000 plates/meter. This might be designated as an ultrahigh resolution LC system. A "good column" (or, perhaps, a high resolution column) generally has reduced plate heights from 2 to 3.5 near the optimum linear velocity (4). This would correspond to a lower value for the high resolution range (10 μm particles, h = 3.5, 25 cm long column) of 7100 plates per column or 29,000 plates/meter.

In SFC either capillary or packed columns may be used. Consider first a comparison between packed column LC and SFC systems. Experimentally obtained (5) plots of the HETP as a function of mobile phase linear velocity for a packed column are shown in Figure 1a. In the LC operating mode a mixture of acetonitrile and water is the mobile phase and in the SFC mode carbon dioxide is the mobile phase. The solvent powers were adjusted in each operating mode so that the capacity factor of pyrene was about the same. In one case the adjustment involved the acetonitrile/water ratio and in the other, the carbon dioxide density.

The first notable point is that the minimum plate heights are the same. This is reasonable and predictable since the minimum plate height ($HETP_{min}$) in a van Deempter curve for a packed column is a function* of the packing particle diameter (d_p) and the analyte capacity factor (k'):

$$HETP_{min} = f(d_p, k')$$

---

*Assuming the simplest case of no eddy diffusion coupling (the coupling approach (6) describes velocity inequalities from lateral diffusion and classical eddy diffusion) so that the "A" term is independent of the linear velocity ($\bar{u}$) and also assuming no resistance to mass transfer in the thin liquid film (2) of the stationary phase, the van Deempter relationship for a packed column is

$$HETP = A + B/\bar{u} + C\bar{u} = 2(\lambda)d_p + 2\ D_{12}/\bar{u} + d_p^2\ \psi\ \bar{u}/D_{12}$$

where HETP, d_p, D_{12}, and k' are defined in the text, λ is the eddy diffusion coefficient, and $\psi = (1 + 6k' + 11k'^2)/(24)(1 + k')^2$.

Then, $\bar{u}_{opt} = \sqrt{2}\ D_{12}/d_p\,\psi^{\frac{1}{2}}$ and $HETP_{min} = 2\ d_p\,\lambda + 2\sqrt{2}d_p\,\psi^{\frac{1}{2}}$.

where $\bar{u}_{opt}$ is the optimum linear velocity corresponding to $HETP_{min}$.

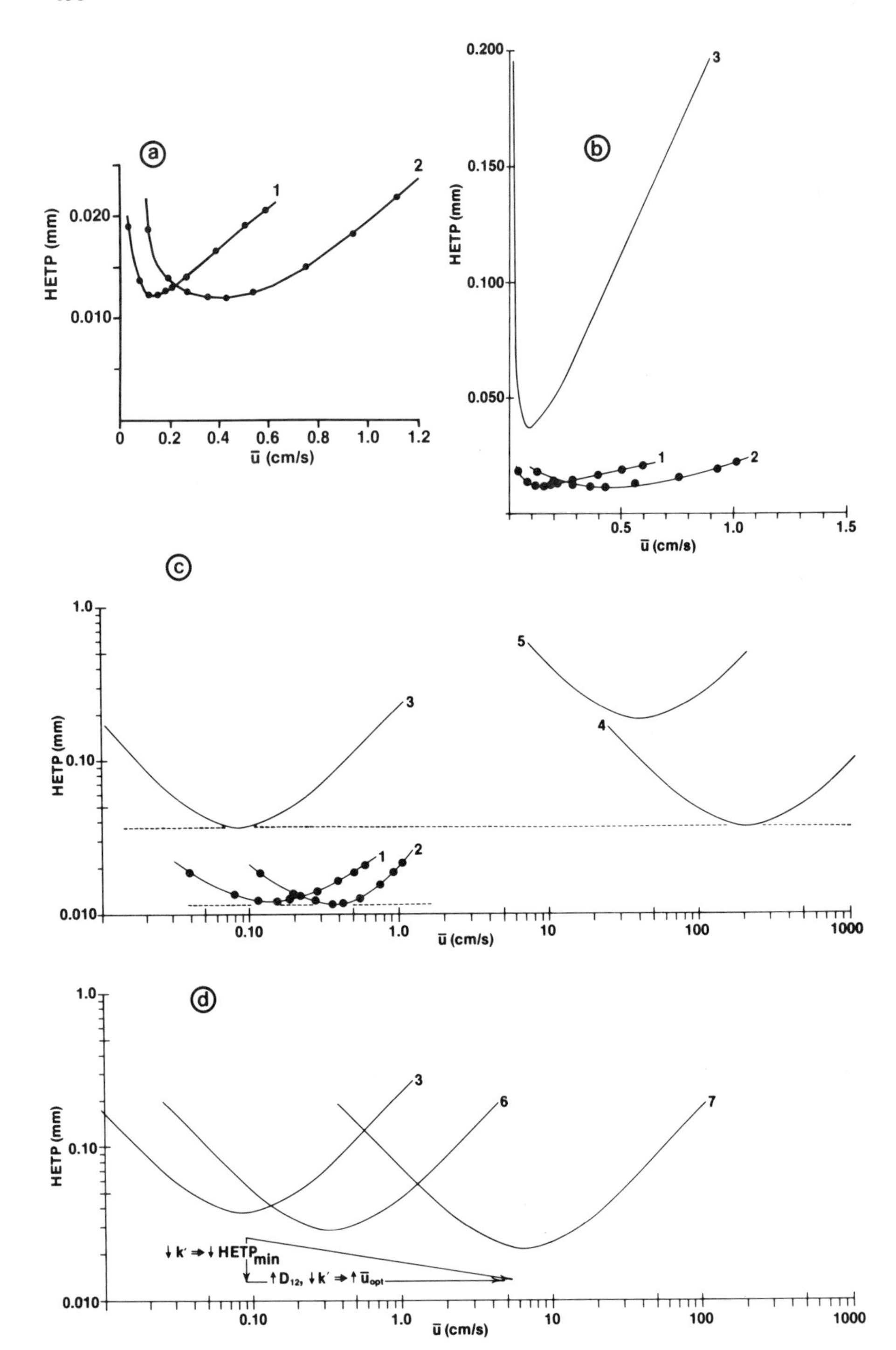

a
HETP (mm)
0.020
0.010
0
0.2
0.4
0.6
0.8
1.0
1.2
ū (cm/s)
1
2
b
0.200
0.150
0.100
0.050
0.5
1.0
1.5
3
c
1.0
0.10
0.010
10
100
1000
4
5
d
6
7
↓ k′ ⇒ ↓ HETPmin
↑ D12, ↓ k′ ⇒ ↑ ūopt

Figure 1. Van Deempter curves for packed column HPLC (#1), packed column SFC (#2), capillary column SFC (#3, #6, #7), and capillary column GC (#4, #5). Curves 1 and 2, experimental data (5); Curves 3 - 7, calculated (see text).

#1 HPLC -- pyrene; Hypersil ODS, 5 µm, 10 cm x 4.6 mm
$k' = 2.85$
CH_3CN/H_2O (70/30), 40°C; take-up solvent, mobile phase

#2 SFC -- pyrene; Hypersil ODS, 5 µm, 10 cm x 4.6 mm
$k' = 2.30$
CO_2, 0.8 g/mL, 40°C; take-up solvent, pentane

#3 SFC -- calculated for pyrene; $d_c = 50$ µm
$k' = 2.30$
CO_2, 0.8 g/mL, 40°C
$D_{12} = 0.00008$ cm^2/s, obtained from Curve 2

#4 GC -- calculated; $d_c = 50$ µm
$k' = 2.30$
D_{12}, assumed 0.2 cm^2/s (0.01 - 1 cm^2/s (2))

#5 GC -- calculated; $d_c = 250$ µm
$k' = 2.30$
D_{12}, assumed 0.2 cm^2/s

#6 SFC -- calculated for naphthalene; $d_c = 50$ µm
$k' = 0.8$
CO_2, 0.3 g/mL, 40°C
$D_{12} = 0.000245$ cm^2/s (13)

#7 SFC -- calculated for pyrene; $d_c = 50$ µm
$k' = 0.32$ (14)
$n\text{-}C_5H_{12}$, 0.08 g/mL, 210°C
$D_{12} = 0.0036$ cm^2/s (14)

Therefore, any column that is packed well for ultrahigh resolution LC separations does not lose its maximum efficiency in SFC separations. However, the linear velocity ($\bar{u}$) must be significantly higher, by a factor of at least three, to obtain the most efficient operation. Again, this is expected since the optimum linear velocity is inversely proportional to the particle diameter and directly proportional to the mobile phase/solute binary diffusivity (D_{12}) and the solute capacity factor (k'):

$$\bar{u}_{opt} = f(1/d_p, D_{12}, k')$$

If the efficiency per unit time, or the resolution per unit time, is considered instead of the overall efficiency, then it can be said that the SFC operating mode is preferable to the LC mode.

The experimental curves in Figure 1a can be used to calculate D_{12} for pyrene in carbon dioxide at the cited temperature and mobile phase density: $D_{12} = 0.00008$ cm^2/s, a value that is quite reasonable (7). This value and the Golay equation* (HETP = $B/\bar{u}$ + $C\bar{u}$) for an open tubular column can be used to generate a typical van Deempter curve for capillary SFC, as shown in Figure 1b for a column inner radius, r_c, of 25 μm . The experimental data from Figure 1a have been plotted again in Figure 1b for direct comparison of the curve shapes, the minimum values of HETP, and the optimum linear velocities. Of course for capillary columns,

$$HETP_{min} = f(r_c, k')$$

and

$$\bar{u}_{opt} = f(1/r_c, D_{12}, k')$$

Typical capillary column GC curves can be similarly generated, as shown in Figure 1c. In this case an intermediate value for the binary diffusivity, 0.2 cm^2/s, was used along with two column inner diameters,** 250 μm and 50 μm. The 50 μm diameter is unusually small for ultrahigh resolution capillary GC work while the 250 μm diameter is in the middle of the diameter range commonly used in this area. As seen in Figure 1c, a comparison of the 50 μm diameter SFC curve to the 50 μm GC curve (Curves 3 and 4) at the same k' emphasizes that the minimum HETP possible in each case is the same (just as for the packed column LC/SFC comparison). Thus, the efficiencies as measured by total plates or plates/meter are equivalent. However, higher optimum mobile phase linear velocities are possible for capillary GC (because of the higher diffusivity) so that capillary GC is superior to capillary SFC in terms of efficiency per unit time. In order to present an order-of-magnitude comparison of all the possible techniques under consideration, a log/log coordinate system was used for Figure 1c. Hence, the curve shapes appear more shallow compared to conventional linear plots.

The statement that supercritical fluids are a hybrid of gases and liquids as we normally encounter them is graphically represented by the series of theoretical capillary column SFC

---

*For a capillary column, the "A" term is zero and the column inner radius, r_c , replaces the particle diameter in the van Deempter equation, resulting in what is known as the Golay equation, so that

$$\bar{u}_{opt} = \sqrt{2}\, D_{12}/r_c\, \psi^{1/2} \quad \text{and} \quad HETP_{min} = 2\sqrt{2}\, r_c\, \psi^{1/2}$$

**These calculations ignore the fact that for small column diameters the resultant significant pressure gradients require a more fundamental form of the Golay equation (8,9). Because the optimum linear velocity becomes increasingly more independent of the diameter as the number of plates increases for diameters of this size (8), there is a trade-off between operating with large numbers of plates but at linear velocities comparable to larger diameter columns and fewer numbers of plates but at increased linear velocities.

curves in Figure 1d. Again, the simplest form of the Golay equation was used to generate these curves. Packed column SFC van Deempter curves would lie at higher optimum linear velocities just as the relative positions of Curves 2 and 3 in Figure 1c just by virtue of the difference between the packing particle diameter and the capillary inner radius (5 μm vs 25 μm).

The purpose of Figure 1d is to emphasize the wide range optimum linear velocities possible with supercritical fluids as the mobile phase. It is important to bear in mind that the optimum linear velocity is a function of k' and D_{12}. Furthermore, k' is a function of the mobile phase solvent power (itself a function of density and chemical composition (χ) with supercritical fluids) and the temperature, and the binary diffusivity is a function of density and temperature:

$$k' = f(\rho, \chi, T)$$

and

$$D_{12} = f(\rho, T)$$

It is instructive to explore these parameters for each of the curves in Figure 1d.

Curve 3. In this case the density of the solvent carbon dioxide is 0.8 g/mL, the temperature is 40°C, D_{12} is 0.00008 cm^2/s, and k' is 2.3--all parameters chosen to make direct comparison between packed (experimental data) and capillary columns. The density of 0.8 g/mL, while at the high end of the 0.1 to 1.0 g/mL range commonly associated with supercritical carbon dioxide as a solvent (10), is a typical density for dissolving larger aromatic compounds (11, 12). (For example, Bowman (11) has shown that the threshold density* of carbon dioxide is 0.1 g/mL for dissolving naphthalene, 0.25 g/mL for dissolving anthracene, 0.5 g/mL for dissolving tetracene, and 0.9 g/mL for dissolving pentacene at 40°C. Gere (12) has presented chromatographic data showing that at 35°C the capacity factor for coronene on a packed reversed phase column ranges from k' = 100 at a carbon dioxide density of 0.8 g/mL down to k' = 35 for carbon dioxide at 1.0 g/mL.)

Curve 6. The values used to calculate this curve were ρ = 0.3 g/mL, D_{12} = 0.000245 cm^2/s, and k' = 0.8, all for carbon dioxide at 40°C. The binary diffusivity is from a range of experimental values measured by Feist and Schneider (13) for

---

*The threshold density is that density of solvent gas at which a solute begins to dissolve in a supercritical fluid (or dense gas) at levels that are detectable (11). In Bowman's work it was UV detection using a mercury lamp; for aromatic compounds the solute solubilities corresponding to threshold densities of carbon dioxide were on the order of micrograms-to-nanograms per milliliter of carbon dioxide.

naphthalene in carbon dioxide at 40°C. The k' for naphthalene is a value determined experimentally on the same type of reversed phase packed column as Curves 1 and 2 at the cited average column density, so its use here is analogous to using k' = 2.3 for pyrene from Curve 2 to Curve 3.

When Curves 3 and 6 are compared in Figure 1d, it is apparent that the minimum HETP lies lower, as expected for a lower k'. The optimum linear velocity is higher in Curve 6 than in Curve 3 not only because the diffusivity is higher but also, to a lesser extent, because k' has decreased. The decrease in k' is probably due to two reasons: 1. naphthalene is more soluble in carbon dioxide than pyrene (as evidenced by threshold densities) and 2. at 40°C naphthalene is fairly volatile, having a vapor pressure of almost one Torr at atmospheric conditions.

Curve 7. The values to calculate this curve were taken from capillary SFC work by Peaden et al (14) who presented van Deempter curves obtained with pyrene in n-pentane at 210°C and 29 atm for several capillary column diameters. A value for D_{12} for pyrene in n-pentane was calculated from one of those curves (for a 300 μm diameter column and a k' of 0.32) in just the same manner as described above for Curve 2 yielding D_{12} = 0.0036 cm^2/s.* The density of n-pentane, estimated from experimental and reduced** pressures and temperatures, is about 0.08 g/mL. (From critical values of 33.3 atm and 469.8 K (15) the reduced pressure is 0.87 while the reduced temperature is 1.028, corresponding to a reduced density of about 0.35 by corresponding states. Hence, this mobile phase is possibly better referred to as a dense gas: i.e., for reduced pressures > 0.7, gas densities rapidly approach liquid densities (16).)

For Curve 7 the minimum value of the HETP is lower than for Curves 3 and 6 (lowest k'), and the optimum linear velocity occurs at even higher values (lowest k' and highest D_{12}, with the effect of the increased diffusivity being far greater). In all three curves the reduced temperature is about the same; for Curves 6 and 7 the solute volatilities at the respective temperatures are probably about the same. Since the maximum solvent powers of carbon dioxide and pentane are similar (17, 18), the density-related volatilization of pyrene into the pentane mobile phase at such a low density is far less significant than the high operating temperature. This is substantiated by the fact that pyrene is

---

* This value appears reasonable when compared to a self-diffusivity value very roughly estimated using a graphical presentation of $(P \cdot D)/(P \cdot D)^o$ as a function of reduced pressure at various reduced temperatures in the manner described by Bird, Stewart, and Lightfoot (15). P is the pressure, D is the self-diffusivity, $(P \cdot D)$ is their product at some high pressure and $(P \cdot D)^o$ is their product at some low reference pressure.

**Reduced parameters: Actual value divided by critical point value.

easily analyzed by GC at temperatures between 200-250°C. Another example of the effect of temperature is the coronene data of Gere (12): with a constant density of 0.8 g/mL (carbon dioxide) an increase in operating temperature from 35°C to 100°C causes a decrease in k' from 100 to 7.9. (Temperatures in excess of 250°C are needed to analyze coronene by capillary GC.) Thus, with these two examples it can be seen that operating temperature is very important. For thermally stable coumpounds, analyses can be done faster, with some improvement in efficiency, at higher operating temperatures. For such analytes, the chromatographer may even wish to use mobile phases with higher critical temperatures so that system pressures are lowered. Other discussions about the simultaneous effects of temperature and density may be found in the literature (19, 20, 21).

What Figure 1d demonstrates is that with a supercritical fluid mobile phase, the chromatographic conditions will range from those of liquid chromatography to those of high pressure gas chromatography. Therefore, the selection of the column and chromatographic conditions in SFC involves the original set of considerations: resolution, resolution per unit time, sample capacity, mobile phase solvent power, and analyte volatility and thermolability. Various approaches to the evaluation of efficiency (i.e., resolution) for SFC as well as LC and GC are presented in Table I where it can be seen that SFC, with packed or capillary columns, compares favorably with the alternatives of packed column HPLC and capillary column GC techniques. In light of that conclusion, it is reasonable to explore the parameter of solvent strength in SFC. In those cases where analytes are involatile and/or thermally labile, a range of relatively high mobile phase densities are required at mild temperatures. When the maximum solvent power of a chosen supercritical fluid is not enough, then that parameter must be changed by changing the chemical identity, or functionality, of the solvent. The remainder of this paper will deal with preliminary experiments that explore the solvent power of carbon dioxide/modifier mixtures. While the work that will be described was pursued for packed column SFC, it is certainly applicable to capillary SFC as well.

PROPOSED FRAMEWORK FOR EXPLORING MOBILE PHASE SOLVENT POWER AND SELECTIVITY

Comparison of Supercritical Fluid Solvents to Conventional Liquid Solvents

As stated in the introduction, the variable solvent power of a supercritical fluid is a nearly linear function of the density of the fluid. It is instructive to compare this way of selecting a solvent power to the more familiar operation used with ordinary liquid solvents. Figure 2 is a schematic of solute solubility as a function of solvent power where the solvent power ranges from that of non-polar hexane to polar water; a multitude of solvents lies in between, only a few of which are shown. Consider a hypo-

Table I. Order-of-Magnitude Comparison of Calculations for Capillary Column GC, Capillary Column SFC, Packed Column SFC, and Packed Column LC for Typical High Resolution Dimensions (d_c = 250 μm, d_p = 5 μm) and Best-Case Dimensions (d_c = 50 μm, d_p = 3 μm). (Experimental values in italics.)

	d_c, d_p (μm)	D_{12}[a] (cm^2/s)	k'	Column Length - L -	$HETP_{min}$[b] (mm)	$\bar{u}_{opt}$[a,c] (cm/s)	t_0[d] (min)	t_R[d] (min)	Total Plates/ Column - n -	Effective Plates/ Column[e] - N -	Effective Plates/ Meter - N/L -	Effective Plates/ Second - N/t_R -
GC	250	0.2	2.3	50 m	0.187	42.8	1.95	6.4	267,000	130,000	2600	340
	50[f]	0.2	2.3	50 m	0.037	214	0.39	1.3	1,351,000	657,000	13,140	8400
SFC	250	0.00008	2.3	50 m	0.187	0.017	4900	16.170	267,000	130,000	2600	0.13
	50	0.00008	2.3	50 m	0.037	0.086	970	3200	1,351,000	657,000	13.140	3.4
	5	*0.00008*	*2.3*	*10 cm*	*0.012*	*0.41*	*0.41*	*1.34*	*8,333*	*4,040*	*40,400*	*50*
	3	0.00008	2.3	10 cm	0.006[g]	0.71	0.23	0.77	16,667	8,100	81,000	175
LC	*5*	*0.000024*	*2.85*	*10 cm*	*0.012*	*0.125*	*1.33*	*5.12*	*8,333*	*4,570*	*45,700*	*15*
	3	0.000024	2.3	10 cm	0.006[g]	0.213	0.78	2.58	16,667	8,100	81,000	52

[a] *For calculated GC values, the intermediate value of 0.2 cm^2/sec for D_{12} was used. (The common range for D_{12} is 0.01-1 cm^2/sec (2). The value of 0.2 cm^2/s seems reasonable in light of values of D_{12} calculated from van Deempter curves in Ref. 2, Figure 1-4, where heptadecane was the compound of interest. Also approximations of D_{12} for pyrene using data for benzene/helium (22)--i.e., assuming the size of pyrene is 4 times that of benzene--indicate this is not unreasonably low, particularly since the LC and SFC curves are for a heavier solute, pyrene.) For SFC and LC, values from experimental van Deempter curves in Figure 1 were used for all D_{12}.*

[b] $HETP_{min} = 2\sqrt{2}\, r_c\, \psi^{1/3}$ *or* $2\lambda d_p + 2\sqrt{2}\, d_p\, \psi^{1/3}$ *; see text for further details.*

[c] $\bar{u}_{opt} = \sqrt{2}\, D_{12}/r_c\, \psi^{1/3}$ *or* $\sqrt{2}\, D_{12}/d_p\, \psi^{1/3}$ *; see text for further details.*

[d] $t_0 = L/\bar{u}_{opt}$ *and* $t_R = t_0(1 + k')$ *where k' is the capacity factor.*

[e] $N = (n)(k'/(k' + 1))^2$

[f] *The numbers here are unrealistically optimistic since the calculations ignored the large pressure drop in a very small capillary column and its effect on the diffusivity. Rigorous calculations would have given a slightly higher $HETP_{min}$ and a lower $\bar{u}_{opt}$ (8,9). For columns with plate numbers greater than 100,000, $\bar{u}_{opt}$ becomes increasingly independent of the column inner diameter. For columns of 30-50 μm inner diameter and up to 9 m long, over 150,000 plates are possible (23).*

[g] *From reduced plate height (h) where $h = HETP/d_p$ so $h_{min} = HETP_{min}/d_p \sim 2$.*

thetical mixture of A, B, and C where only A is significantly soluble in hexane, A and B are soluble in a 50/50 mixture of hexane and ether, and all three components are soluble in ether. These mixture components can be separated from each other by varying the solvent power from that of hexane to that of ether--i.e., the chemical identity of the solvent must be changed.

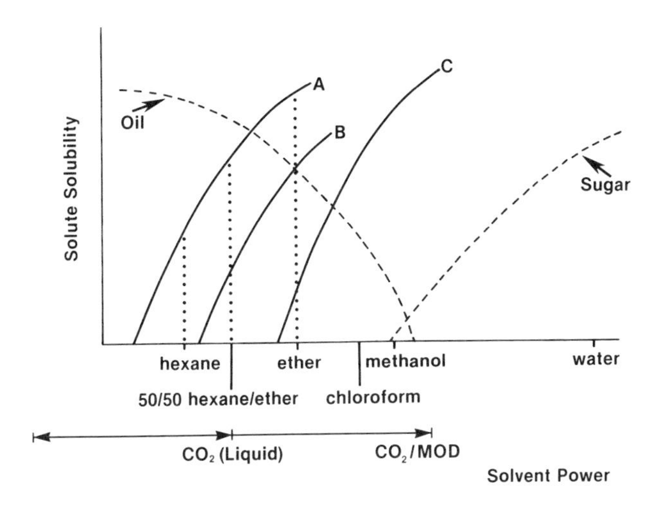

Figure 2. Generalized behavior of solute solubility as a function of solvent power. Solvent placement is approximately according to the Snyder polarity ranking and the curves of oil and sugar are based upon various published data.

Another way to effect the separation is to use supercritical carbon dioxide. As indicated below the x-axis (Figure 2), the solvent power of carbon dioxide ranges from low values at low fluid densities to a maximum value, about that of liquid carbon dioxide, at higher densities. Here the chemical identity of the fluid remains the same; only the system pressure is changed. If the maximum solvent power is not sufficient for a particular application, there are two choices: 1. change the identity of the supercritical fluid (e.g., to supercritical acetone, methanol, ammonia--often with a concomittant large change in operating temperature because of higher critical temperatures) or 2. use mixtures of carbon dioxide and other more polar solvents. The most effective way to make this choice routinely with minimal trial-and-error experimentation is to evaluate the solvent power of carbon dioxide with reference to other solvents. This allows the use of existing solute solubility information available for solvents determined to be similar to carbon dioxide. A semi-empirical approach provided just an estimate in 1968 (24).

More recently, two other approaches have been undertaken (17, 18). They were the calculation of the Hildebrand solubility parameter as a function of density using tabulated thermodynamic data for carbon dioxide and Raman spectroscopy of test solutes dissolved in supercritical carbon dioxide compared to liquid solvents to evaluate solvent-solute interactions. The results of these recent approaches indicated that while the maximum solvent power of carbon dioxide is similar to that of hexane, probably somewhat higher, there is some solvent-solute interaction not found with hexane as the solvent. The limiting solvent power of carbon dioxide is resolved by choosing the alternative of a supercritical fluid mixture as the mobile phase. The component added to the supercritical fluid to increase its solvent power and/or to alter the chromatograph column is referred to as the "modifier."

As in HPLC, the problem is now what to use as the second component in the mobile phase solvent mixture. There are many organic solvents that are soluble in liquid (25, 26, 27) and supercritical (27, 28) carbon dioxide and so the chromatographer is presented with numerous possibilities from which to choose. In a way, in SFC there is an even greater range of modifier choice than available in conventional HPLC because the SFC instrumentation allows pumping of highly compressible liquids with low boiling points, with carbon dioxide being an extreme example. Of course, detector-compatibility, toxicity, expense, and purity are modifier considerations that still apply in SFC as well as in HPLC.

The Snyder solvent classification scheme (29, 30) that has been successfully applied to LC systems (31-37) for LC separation optimization appears to be a general framework that can be used for guidance in the selection of the modifier. Some preliminary experiments have been done to determine the applicability of this approach to SFC systems as a prelude to a more in-depth, statistically-designed modifier selection study. These first experiments are the topic of this paper; the basis of the proposed modifier selection scheme will now be described in detail.

Proposed Framework

Snyder has discussed liquid solvent characterization on several occasions (29,30,38,39). One of the very interesting points is that when several solvents have essentially the same solvent power as described by the Hildebrand solubility parameter (40), those solvents often do not dissolve a particular solute to the same extent. This is attributed to the fact that the Hildebrand solubility parameter does not, and cannot by regular solution theory, reflect solute-solvent interactions stronger than dispersive type forces. The Hildebrand solubility parameter for a solvent, the square root of the molecular cohesive energy per unit volume, is obtained from thermodynamic properties of only the solvent--i.e.,

$$\delta = \left[(\Delta H^V - RT)/V \right]^{1/2}$$

where δ is the Hildebrand solubility parameter, ΔH^V is the heat of vaporization, R is the gas constant, T is the temperature, and V is the molar volume.

Snyder has derived (29) and refined (30) a solvent classification scheme, using data of Rohrschneider (41), in which each solvent is assigned an overall "polarity," P'. (The polarity is also referred to as the polarity index or the chromatographic strength.) The polarity is an empirical ranking of solvents and ranges from a value of 0.1 for hexane to 10.2 for water. This measure of solvent power, similar to the Hildebrand solubility parameter, is derived from the interaction of each solvent with a set of test solutes (ethanol, dioxane, nitromethane) that probe three possible types of solvent interactions with the solute: proton acceptor, proton donor, and strong dipole. Hence, the overall polarity, P', is the sum of that part of the solvent polarity due to each type of solvent-solute interaction:

$$P' = P'_{\text{proton acceptor}} + P'_{\text{proton donor}} + P'_{\text{strong dipole}}$$

Solvents can be compared in terms of their selectivities by constructing tricoordinate plots of the three selectivity fractions. The selectivity fraction of each of the three interactions can be defined as that amount of the solvent polarity due to the particular interaction divided by the total polarity. Then each selectivity fraction ranges from 0 to 1 -- e.g., from no proton acceptor character to pure proton acceptor character. Snyder presents selectivity fractions for 80 solvents (30); the placement of a few of those solvents within the selectivity triangle is shown in Figure 3. Quite interestingly the 80 solvents under consideration fell into eight distinct selectivity groups--e.g., alcohols grouped together; ethers were nearby but distinct; and "universal" solvents like 2-methoxyethanol, dimethylsulfoxide, and dioxane were centered within the triangle.

It should be emphasized that the selectivity mapping shown in Figure 3 says nothing about the overall polarity (solvent strength) of each solvent. In the overall polarity ranking, toluene with a P' of 2.4 lies significantly lower than dimethylsulfoxide with a P' of 7.2. The total classification scheme is summarized in Figure 4, where the overall polarity P' increases along a linear vertical axis, piercing selectivity triangles.

When it is desired to study a wide range of solvent selectivities, only three solvents--each from a selectivity triangle vertex--are needed, not the huge numbers that are possible. Any solvent whose selectivity lies within a chosen selectivity

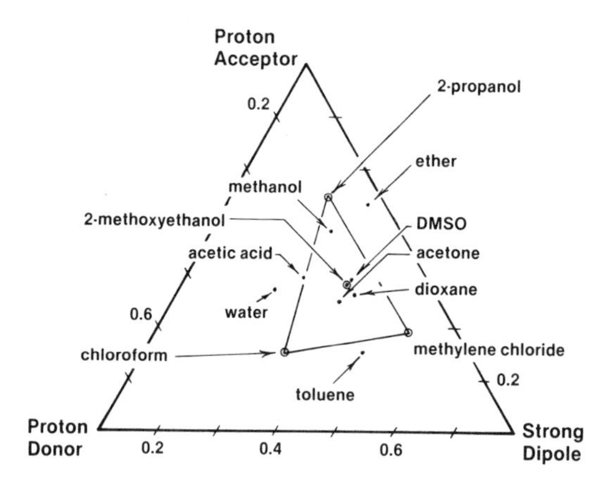

Figure 3. Selectivity mapping for typical liquid solvents using selectivity fraction values tabulated by Snyder (30).

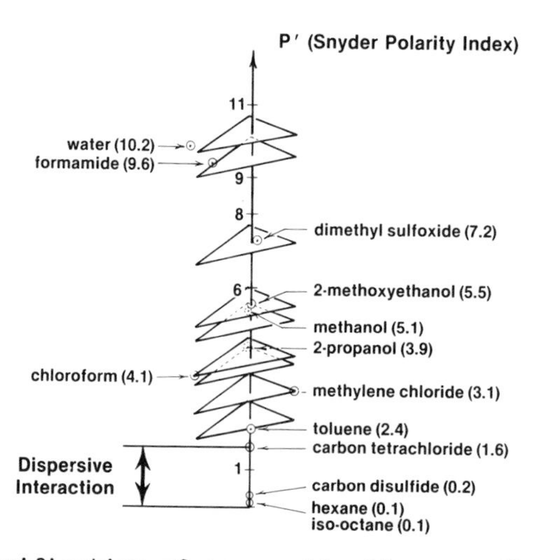

Figure 4. Classification of common liquids according to overall polarity (Snyder Polarity Index) and solvent selectivity. Overall P' values and solvent selectivity fractions are those tabulated by Snyder (30).

triangle can be approximated by the proper combination of the three vertex solvents in a diluent--i.e., the desired selectivity is achieved by the proper ternary combination and the desired overall polarity, by the proper quaternary combination (the selectivity ternary mixture in a diluent fourth component).

Within the last several years HPLC separations have been optimized in terms of the most appropriate mobile phase composition for a particular set of solutes by exploring the whole plane of solvent selectivities using this solvent classification scheme with a minimal number of measurements in statistically-designed experiments. For reversed phase HPLC systems, the selectivity triangle is often defined by methanol, acetonitrile, and tetrahydrofuran with water as the diluent (37).

However, in normal phase adsorption systems (or liquid-solid chromatography) the interaction of the mobile phase solvent with the solute is often less important than the competing interactions of the mobile phase solvent and the solute with the stationary phase adsorption sites. Solute retention is based upon a displacement mechanism. Multicomponent mobile phases and their combination to optimize separations in liquid-solid chromatography have been studied in detail (31-35). Here, solvents are classified as to their interaction with the adsorption surface (Reference 32, in particular):

1. non-localizing solvents that are non-polar to intermediate polarity. The interaction with the surface ranges from dispersive to polar but any part of the surface is used for interaction; the solvent is not localized on an active site. Solute retention is a function of solvent strength (ε^0), simply related to the mole fraction of each component in the mobile phase.

2. strong dipolar, localizing solvents. These mobile phase solvents interact strongly with adsorption sites; their solvent strengths depend upon the mole fraction of the solvent on the stationary phase. When the active sites of the adsorbent surface are covered by the localizing solvent molecules, further interaction of the solvent molecules with the surface is non-localizing. When localizing solutes and solvents are present, an interaction between the two effects results in a cross term between the relative localization of the solute and the solvent localization and its surface coverage. Strong dipolar, localizing solvents compared at about the same solvent strength and the same solvent parameter (32) exhibit no solvent-specific localization--i.e., the solute capacity factor (k') and separation ratio (α) are functions only of the mobile phase solvent parameter,

3. basic, localizing solvents. A solvent-specific localization interaction exists for these solvents; solute capacity factors and separation ratios are functions of the solvent identity as well as the mobile phase solvent parameter.

Therefore, in liquid-solid chromatographic systems selectivity in a separation is determined by the mobile phase solvent strength and interaction with the adsorbent. The general selectivity triangle for such systems is shown in Figure 5a; in Figure 5b, examples of non-localizing, non-solvent-specific localizing, and solvent-specific localizing solvents are shown as well as the actual liquid-solid selectivity triangle solvents heretofore used. (31)

While remembering to keep the different separation mechanisms and their different solvent classification schemes in mind, the chromatographer in SFC systems should be able to use such selectivity triangles for appropriate choices of modifier. For SFC experiments with bonded phase columns, methylene chloride as the large dipole moment, chloroform as the proton donor, and 2-propanol as the proton acceptor were chosen as the selectivity ternary solvents. This selectivity triangle is compared to that commonly used (37) to optimize reversed phase systems in Figure 5c. The miscibility of carbon dioxide with a large number of solvents allows a large selectivity triangle to be chosen. A center-point selectivity can be achieved by an equal volume mixture of each of the three components (i.e., 1/3 chloroform + 1/3 methylene chloride + 1/3 2-propanol). Since the selectivity fractions of 2-methoxyethanol are very nearly the arithmetic averages of each of the selectivity fractions of the three SFC vertex modifiers and since 2-methoxyethanol has a high overall polarity (P' = 5.5), that solvent was also chosen as a possible modifier to be studied.

Figure 5. Various selectivity triangles.

a. General selectivity triangle for liquid adsorption chromatography

b. Selectivity triangle used for optimization of liquid adsorption chromatographic separations along with examples of localized basic, localized dipolar, and non-localized solvents.

c. Comparison of the selectivity triangle used for the optimization of reversed phase liquid chromatographic separations to that chosen for the supercritical fluid modifier selectivity survey.

d. The two adsorption chromatography selectivity triangles available with chosen SFC modifiers compared to that used in LC systems.

e. Comparison of selectivity triangles used in LC and SFC.

______ Liquid adsorption chromatography
. . . . Liquid reversed phase partition chromatography
- - - - Adsorption and partition chromatography with supercritical fluid mobile phases

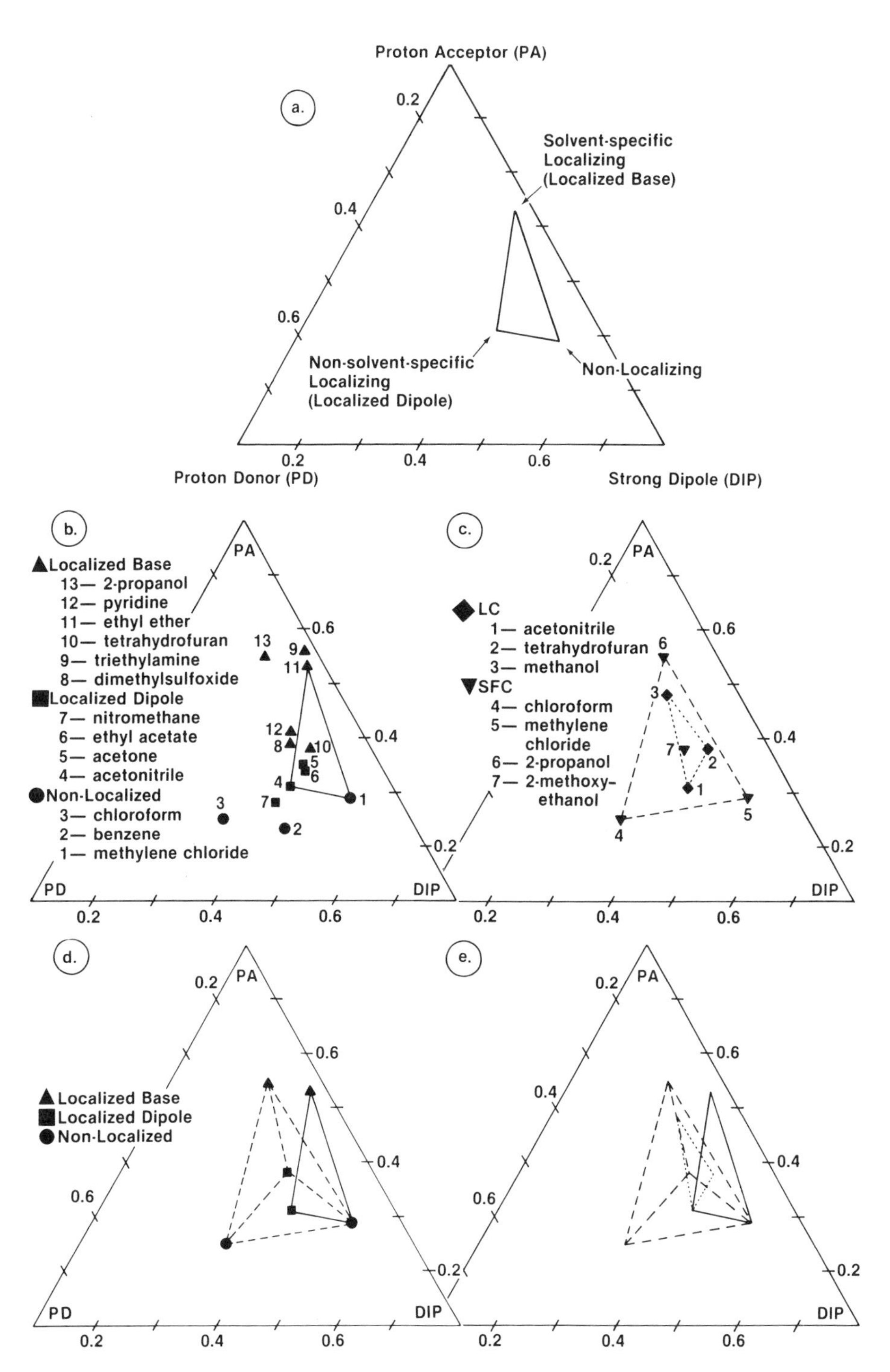
Proton Acceptor (PA)
a.
0.2
0.4
0.6
Solvent-specific Localizing (Localized Base)
Non-solvent-specific Localizing (Localized Dipole)
Non-Localizing
0.2
0.4
0.6
Proton Donor (PD)
Strong Dipole (DIP)
b.
PA
Localized Base
13— 2-propanol
12— pyridine
11— ethyl ether
10— tetrahydrofuran
9— triethylamine
8— dimethylsulfoxide
Localized Dipole
7— nitromethane
6— ethyl acetate
5— acetone
4— acetonitrile
Non-Localized
3— chloroform
2— benzene
1— methylene chloride
PD
DIP
c.
PA
LC
1— acetonitrile
2— tetrahydrofuran
3— methanol
SFC
4— chloroform
5— methylene chloride
6— 2-propanol
7— 2-methoxy-ethanol
d.
Localized Base
Localized Dipole
Non-Localized
e.

For adsorption chromatography on silica* with a supercritical mobile phase, the selectivity triangle vertex solvents would be methylene chloride or chloroform as the intermediate polarity non-localizing solvent, 2-methoxyethanol as the non-solvent-specific localizing solvent,** and 2-propanol as the solvent-specific localizing solvent. Therefore, the SFC modifier selectivity triangle chosen for partitioning stationary phases encompasses two selectivity triangles for SFC adsorption chromatography (Figure 5d). Finally, all selectivity triangles discussed here are compared in Figure 5e.

In these SFC studies carbon dioxide is the diluent, analogous to the use of hexane in LC studies and reasonable in light of the results presented in References 17 and 18 and of exploratory chromatographic experiments--all showing that while carbon doxide may have slight selectivity (towards the strong dipole vertex, perhaps due to its quadrapole moment), it has an overall maximum polarity of about that of hexane.

RESULTS AND DISCUSSION

The purpose of these first experiments in the modifier survey work was to determine the applicability of the Snyder solvent classification scheme to the choice of modifier to see if an in-depth study would be warranted. The experimental conditions are described in the final section of this paper.

The initial experiments at the vertices of the selectivity triangles did show that large differences in separation could be induced by the modifier choice. It was seen that the capacity factors and separation ratios were not only functions of the modifier identity but also functions of the modifier concentration in the modifier/carbon dioxide mobile phase mixtures. Furthermore, adsorption chromatography with a supercritical fluid mobile phase does exhibit fast column re-equilibration times upon a change in mobile phase composition. However, methylene chloride and chloroform were not acceptable modifiers on the fully active silica adsorbent, although chloroform, in particular, did demonstrate a selectivity different from 2-propanol and 2-methoxyethanol as modifiers. Finally, chromatographic peak distortion, the onset of which appeared to be system-specific, was often but not always observed. Each of these points will now be discussed in more detail followed by a concluding outline of possible experiments suggested by these results.

---

*Snyder and Glajch (32) note that alcohols can be used as solvent-specific localizing solvents for adsorption on silica, with behavior predictable by their model; this is not true for alumina systems (34).

**This solvent appears to lie at the dividing line between non-solvent-specific and solvent-specific localizing solvents; its actual classification has not been demonstrated experimentally.

Modifier Selectivity

Xanthines. It was found that the capacity factors and the separation ratios were both functions of the identity of the modifier as well as of the quantity of the modifier in the carbon dioxide. In Figure 6, it can be seen that with 9.5% 2-methoxyethanol in carbon dioxide* the first two components of a xanthine mixture, caffeine and theophylline, are not well-separated but the elution order is caffeine, theophylline, theobromine, and xanthine. At 6.5% 2-methoxyethanol in carbon dioxide, there is essentially baseline resolution of the first three components in less than one minute; xanthine (not very soluble in liquid 2-methoxyethanol and not at the usual 1 mg/mL sample concentration but at some undetermined saturation concentration) at a k' of over 5 has begun to tail and to disappear into the baseline. While it is obvious that the capacity factors, k', increase with a decrease in modifier concentration from 9.5 to 6.5% 2-methoxyethanol in CO2, less obviously the separation ratios (for example, with respect to caf-

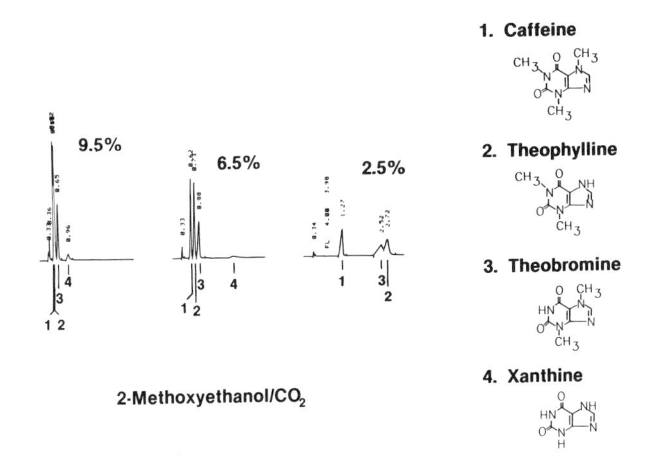

Figure 6. Separation of xanthines at various concentrations of 2-methoxyethanol in carbon dioxide.

---

* In the 1082 LC/SFC flow control system the contribution of the pump B solvent stream to the overall solvent flow, computed from the independent volumetric flow rate measurements, depends upon the relative compressibilities of the two fluids. Here it is assumed that pure carbon dioxide and the 10% (molar) modifier/carbon dioxide mixtures have about the same compressibilities and so the actual modifier concentration is given simply by the product of (%B)(0.1).

feine) also all increase with the change in modifier concentration, and not by a constant factor. A more dramatic change in the separation ratios is shown by the chromatogram of the xanthines with 2.5% 2-methoxyethanol in carbon dioxide where there is a change in elution order to caffeine, theobromine, and then theophylline.

Estimates of efficiencies and resolution can be made by considering the chromatogram for 6.5% 2-methoxyethanol in carbon dioxide. The resolution between caffeine (k' = 0.88) and theophylline (k' = 1.24) is 2.17 and between theophylline (k' = 1.24) and theobromine (k' = 1.7) is 2.00, yielding an average effective number of plates (N) of 884 and an efficiency-per-unit-time parameter of 20 effective plates/s (theophylline). This is somewhat higher than expected for an LC system (Table I) and lower than the most efficient SFC system. Compared to the values in Table I, the linear velocity is higher than optimum, the diffusivity is different, and the analyte capacity factor (elution time) is lower.

The effect of the modifier identity can be seen in Figure 7 which shows the separation of the xanthines with 9.5% 2-propanol in carbon dioxide. Compared to 9.5% 2-methoxyethanol, the capacity factors of the first three components are significantly larger (by a factor of 1.6 to 2.5); the elution order is also

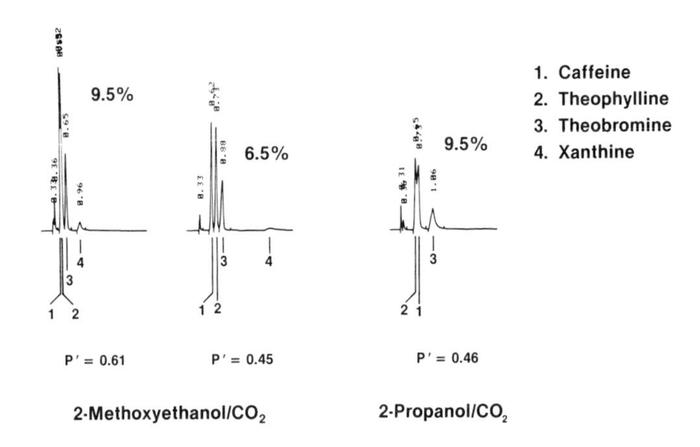

Figure 7. Separation of xanthines with 2-propanol as the modifier compared to 2-methoxyethanol as the modifier.

different, with theophylline preceding caffeine. Since the overall polarity, P', of 2-methoxyethanol is higher than that of 2-propanol (P' = 5.5 vs 3.9, Figure 4), it would more fair to compare the mobile phase mixtures at equal overall polarities. With the assumption that the polarity of carbon dioxide is 0.1, like hexane and pentane, with no significant selectivity, it can be calculated after the manner of Snyder* that 6.5% 2-methoxyethanol in carbon dioxide has about the same overall polarity as 9.5% 2-propanol in carbon dioxide. A comparison of the separations of the mixture using the equal polarity solvents shows that the elution times and order are still distinctly different for the two modifiers.

Further experiments with 2-propanol as the modifier in the xanthine separation showed another elution order change with a modifier concentration change as shown in Figure 8. Another phenomenon, particularly evident in the chromatograms at the lower 2-propanol modifier percentages, is peak distortion. The distortion appears to occur when the mobile phase solvent power becomes marginal for competiting with the column for the solute or for displacing the solute from the column active sites. Its onset seems to be related to the particular solute-mobile phase composition-chromatograph column combination under consideration--i.e., it is specific to each different combination.

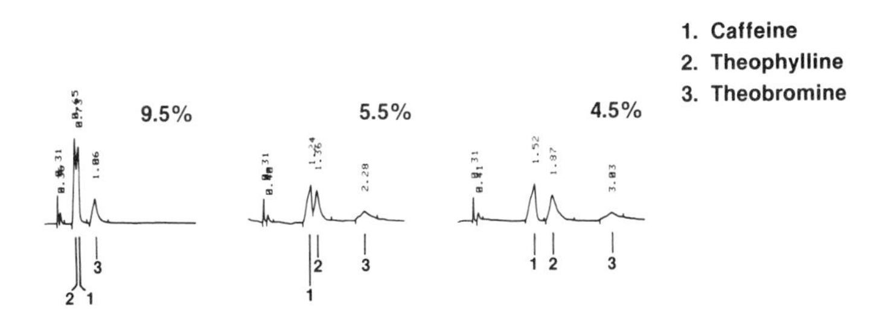

Figure 8. Separation of xanthines at various concentrations of 2-propanol in carbon dioxide.

---

* $P' = \Phi_A P'_A + \Phi_B P'_B$ where Φ_A, Φ_B are volume fractions of components A and B and P'_A, P'_B are overall polarities of A and B (39).

Hormones. Another example of the dependancies of component elution time and order upon the modifier identity and content in modifier/carbon dioxide mixtures is presented in Figure 9. The chromatogram with chloroform as the modifier demonstrates fairly typical chromatographic behavior with methylene chloride and chloroform as modifiers. In general, these two modifiers simply did not yield acceptable chromatograms on the Hypersil SIL column; the reason for this failure will be explored later in this discusson. Even if chromatography with these modifiers appears to be useless at a first glance, there is a different selectivity when using them compared to 2-methoxyethanol and 2-propanol as seen in Figure 9 and as can also be seen in Figure 10.

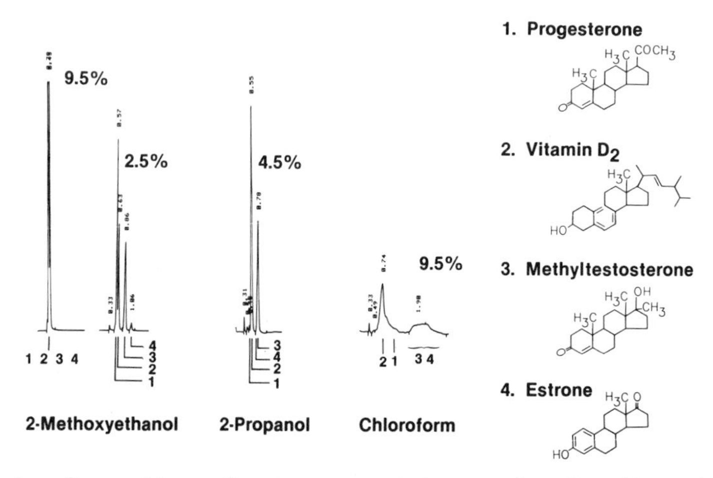

Figure 9. Separation of a hormone mixture using 2-methoxyethanol, 2-propanol, and chloroform as modifiers in carbon dioxide.

Substituted anthraquinones. In the latter figure, 9.5% 2-propanol in carbon dioxide as the mobile phase results in a separation very similar to that with 5.5% 2-methoxyethanol in carbon dioxide. In both cases 1,8-dihydroxyanthraquinone elutes with and right after anthraquinone so those components are not separated, even at much lower modifier concentrations; the retention times of anthraquinone and 1,8-dihydroxyanthraquinone increase together with the tailing of the 1,8-dihydroxyanthraquinone becoming more and more pronounced as the modifier concentration is decreased. However, with chloroform as the modifier, those two components are significantly split apart, with 1-aminoanthraquinone eluting in between them. Another interesting feature of Figure 10 is that 1,2-dihydroxyanthraquinone was never eluted from the Hypersil SIL column, no matter which modifier was used. This difference in isomer elution between the

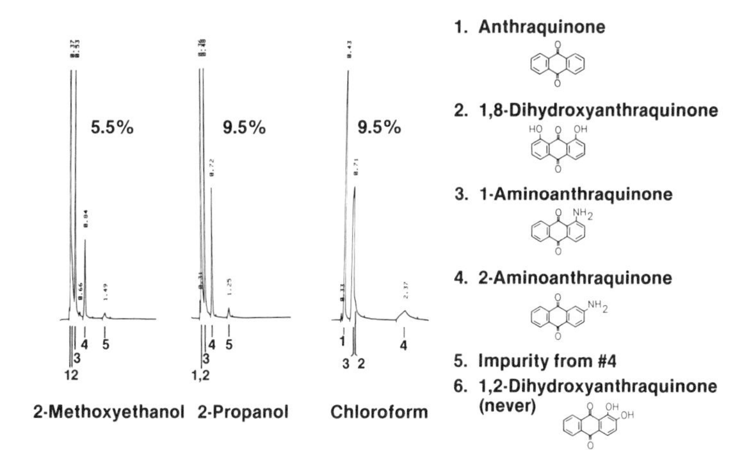

Figure 10. Separation of a mixture of substituted anthraquinones with 2-methoxyethanol, 2-propanol, and chloroform as modifiers in carbon dioxide.

1,8- and the 1,2- dihydroxyanthraquinones seems to fit old open column silica chromatography data for substituted anthraquinones which showed that internal hydrogen bonding between the ketone oxygens and the hydroxy groups on adjacent carbons (the 1,4,5 and 8 positions) could occur and thus interfere with solute adsorption (42). Compounds with hydroxy groups in the other positions that do not allow internal hydrogen bonding are far more strongly adsorbed. The same behavior is observed for amino groups. Finally, amino-substituted anthraquinones are less strongly adsorbed than hydroxy-substituted anthraquinones so the elution of 2-aminoanthraquinone before the 1,2-dihydroxyanthraquinone (specifically the substitution on the number 2 position) also agrees with earlier adsorption chromatography experience (42).

Xanthone, flavone and similar compounds. A difference in chloroform specificity from 2-methoxyethanol is again demonstrated in Figure 11. Xanthydrol is eluted before xanthone and flavone with chloroform as the modifier. Perhaps this shows the useful coupling of a proton donor solvent and proton acceptor solute (a large secondary solvent effect in an adsorption system (43)), the interaction that was hoped for with the selection of chloroform as one of the selectivity triangle modifiers and apparent here because of a less strong adsorption of the xanthydrol on the fully active silica than some of the other basic solutes used in the preliminary studies.

The tailing shown by the Biochanin A chromatographic peak in Figure 11 with 3.5% 2-methoxyethanol in carbon dioxide was typi-

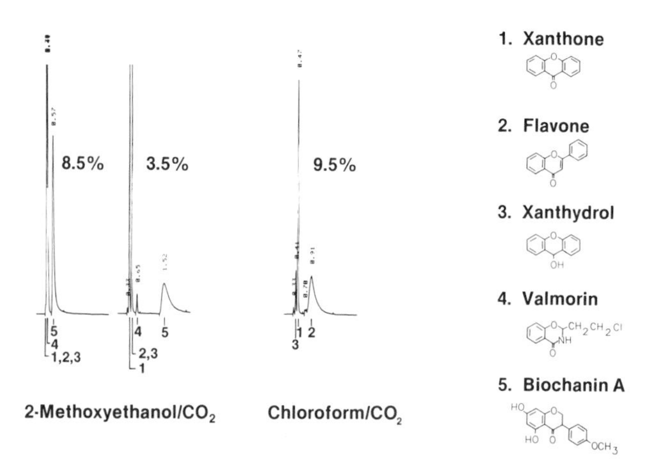

Figure 11. Separation of xanthone, flavone and similar compounds with 2-methoxyethanol and chloroform as modifiers.

cally observed with hydroxy group substitutions on aromatic rings. An increased modifier concentration can help to minimize the tailing, as shown by the chromatogram for 8.5% 2-methoxyethanol in carbon dioxide. It is somewhat surprising to the author that the analgesic drug, Valmorin, does not exhibit tailing or peak distortion at the lower 2-methoxyethanol concentrations; perhaps this is another case of internal hydrogen bonding partially interfering with solute adsorption on the silica.

Table II is a summary of the various solute-modifier combinations that were tried on the Hypersil SIL column. Additional information is presented in Reference 44.

Adsorption Chromatography

Chloroform and methylene chloride as modifiers. In general, even though they demonstrated selectivities different from the other modifiers, chloroform and methylene chloride were disappointing as modifiers for adsorption chromatography on the fully active Hypersil SIL silica column. This may arise more from the fact that the silica column was fully active than from their lower solvent strengths (ε^0) in the silica adsorption eluotropic series. Snyder has emphasized the need for adsorbent deactivation where the deactivating molecules are selectively adsorbed on the strongest sites for decreased surface heterogeneity, increased linear capacity, and high chromatographic efficiencies (43). It is plausible that both of the polar localizing modifiers (2-methoxyethanol as the non-solvent-specific and 2-propanol as the solvent-specific localizing solvents) served as deactivators to the strongest active sites, with essentially irreversible adsorption until column reactivation at elevated temperatures with pure

Table II. Chromatographic Behavior of Various Test Solutes with Modifier/CO_2 Mobile Phases

		2-Methoxyethanol	2-Propanol	Chloroform	Methylene Chloride
I	Anthracene	1	1	1	1
	Anthraquinone	1	1	1	1,2
	Xanthone	1	1	1	2
	Acenaphthenequinone	1,2	1	2	4
II	1-Aminoanthraquinone	1	1	2	2
	2-Aminoanthraquinone	1,2	1	2	6
	1,8-Dihydroxyanthraquinone	3	3	2	6
	1,2-Dihydroxyanthraquinone	5	5	5	6
	Anthraquinone	1	1	1	1,2
III	Caffeine	1,2	1,2	5	5
	Theophylline	1,2	1,2	5	5
	Theobromine	1,2	1,2	5	5
	Xanthine	1,3	5	5	5
	Adenine	1,2	3	5	5
IV	Diphenylphthalate	1	1	1,2	2
	Dimethylterephthalate	1	1	1,2	2
	Di-n-butylphthalate	1	1	1,2	2
	Phthalic acid	5	5	5	5
V	Xanthone	1	1	1	2
	Flavone	1	1	3	5
	Xanthydrol	1,2	1	1	5
	Biochanin A	1,3	3	5	5
	Valmorin	1,2	1	5	5
VI	Progesterone	1	1	4	5
	Methyltestosterone	1,2	1,2	4	5
	Estrone	1,2	1,2	4	5
	Vitamin D_2	1	1	2,4	4
VII	1-Aminoanthraquinone	1	1	2	2
	1-Nitronaphthalene	1	1	1	1
	1-Naphthyl-acetic acid	3	5	5	5
	Acenaphthenequinone	1,2	1	2	4
	p-Nitrophenylacetonitrile	1	1	2	1
	Anisyl alcohol	1	1	4	2
	2-Phenylethanol	1	1	4	5

1 -- Sharp chromatographic peaks; 2 -- Peak distortion at some modifier concentration (100%B to 15%B; the whole range was not necessarily studied for each modifier); 3 -- Tailing; 4 -- Poorly eluted, distinct from the characteristic peak distortion and from tailing; 5 -- Never eluted at maximum modifier concentration; 6 -- Never tried solute/modifier combination.

carbon dioxide, as well as the desired reversible localizing agents at the remaining weaker active sites. Furthermore, it would be expected that the moderately polar non-localizing modifiers, chloroform and methylene chloride, could not function as deactivators. Thus, the strongest active sites were available for chemisorption of the most polar, localizing solutes when chloroform and methylene chloride were used as modifiers.

Fast column re-equilibration. Some limited measurements were made to assess the time necessary for column-mobile phase equilibration when a ballistic change in modifier concentration was made. Thus far, these measurements have involved only 2-methoxy-

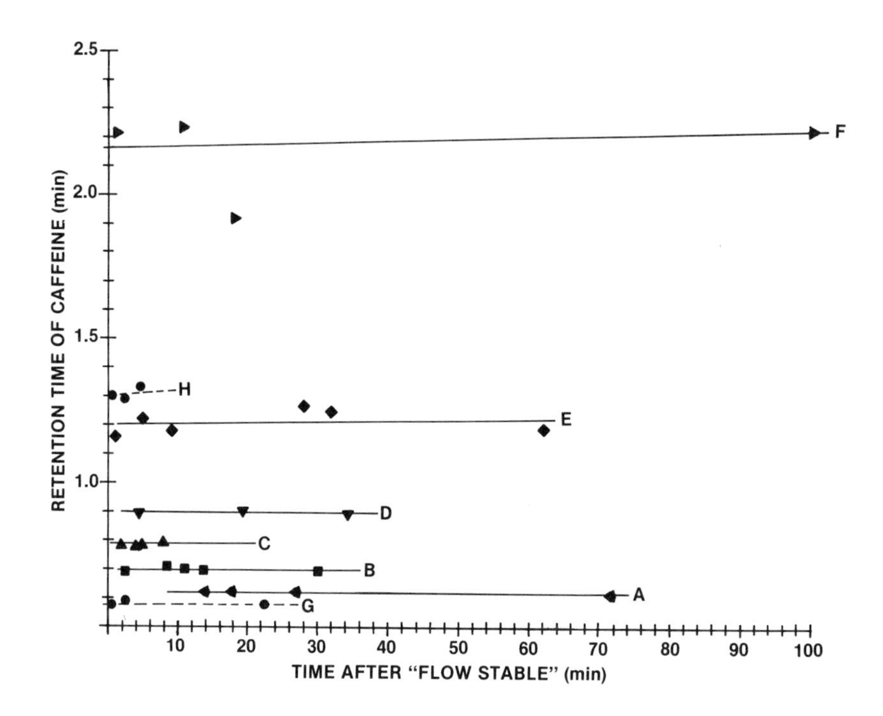

Figure 12. Column re-equilibration upon a ballistic change in mobile phase composition. Curves A - F, the same Hypersil SIL column used for chromatograms in this paper; Curves G and H, another Hypersil SIL column.

A - %B: 75% ==> 65%
B - %B: 65% ==> 55%
C - %B: 55% ==> 45%
D - %B: 45% ==> 35%
E - %B: 35% ==> 25%
F - %B: 25% ==> 15%
G - %B: 25% ==> 65%
H - %B: 65% ==> 25%

ethanol as the modifier and the xanthines as solutes and were carried out as outlined in the experimental section. The results of these experiments are presented in Figure 12, in which the retention time of caffeine is plotted as a function of time elapsed from the "FLOW STABLE" condition. Generally, it can be said that re-equilibration is attained in a matter of 2 to 5 minutes, not only for small changes of 10%B (1% actual modifier concentration) but for larger changes of 40%B (4% actual modifier concentration)--i.e., the retention times of caffeine rapidly reach values bracketed by the variation in retention times after equilibration. Furthermore, the retention time stability at lower modifier concentration often appears less than at higher concentrations, probably due to the onset of peak distortion with its attendant difficulty in peak integration. Certainly, more experiments with different modifiers, different solutes, and a far larger range in modifier concentration are necessary for a fair and general comparison of the short re-equilibration times of SFC to the typically large equilibration times of normal phase LC on the order of hours (45, 46).

Peak Distortion

As noted earlier, peak distortion was commonly observed at lower solvent powers--with a particular "lower" solvent power being specifically associated with the solute of interest. For example, as shown in Figure 6, very sharp, well-shaped chromatographic peaks are obtained for caffeine, theophylline, and theobromine with 6.5% 2-methoxyethanol in carbon dioxide; however, with the lower 2.5% 2-methoxyethanol in carbon dioxide concentration, the onset of distortion is evident. For anthracene on the same Hypersil SIL adsorption column, no distortion is observed with pure carbon dioxide for column midpoint* densities above 0.5 g/mL (60°C, ρ_{in} = 0.60 g/mL, ρ_{out} = 0.34 g/mL, FLOW setting of 5 mL/min); however, even with a constant outlet density, distortion can be observed when the flow rate is slow enough that the column inlet density and thus the density/pressure-drop profile across the column does not provide this minimum midpoint density.

The origin of this distortion has not yet been completely determined. At first column overloading by solute or a poorly packed column were suspected; these possibilities were both eliminated. The first was explored by lowering the injected solute quantity to a few nanograms. In that case the distortion

*Column midpoint density is that density corresponding to the midpoint pressure, $P_{mid} = (P_{in} + P_{out})/2$; ρ_{mid} does not necessarily equal the arithmetic average of the inlet and outlet densities, $(\rho_{in} + \rho_{out})/2$. For the experimental conditions cited in the text giving values of ρ_{mid} of about 0.5 g/mL, ρ_{mid}/ρ_{ave} is about 1.05.

persisted. A second Hypersil SIL column, with an acceptable minimum plate height, was substituted for the first column; however, the peak distortion behavior at low solvent powers was still observed. It has been suggested to the author that the distortion is due to using a highly polar take-up solvent (2-methoxyethanol in this work) that interferes with the actual chromatographic elution by less polar mobile phases for some initial length of the column. Another possibility that has been mentioned is that the injection procedure causes the distortion.

A limited number of experiments were conducted to evaluate the last two possibilities. First, the take-up solvent for two test solutes, anthracene and naphthalene, was cyclopentane (with solution concentrations still about 1 mg/mL and injection aliquots still about 0.3 - 0.5 μL into a 20 μL loop) instead of 2-methoxy-ethanol. The distortion persisted even when a gentle stream of air was used after aliquot deposition in the sampling loop to blow off the take-up solvent. Furthermore, the onset of the distortion for naphthalene occurred at a lower carbon dioxide density than for anthracene. When the column was changed to a partially chemically deactivated column, Hypersil SAS, where short alkyl chains are bonded to active sites, onsets of peak distortion for anthracene* and for naphthalene were shifted to lower carbon dioxide densities (and still different from each other) than for the fully active Hypersil SIL adsorption column. Only slight peak distortion was observed for a Hypersil ODS column at even lower column midpoint densities (0.33 g/mL with the back pressure at 1170 psig, the FLOW setting at 1.90 mL/min, and 40°C; distortion was observed for only anthracene and not naphthalene).

In order to explore the second proposed origin, the injection procedure, the solvent conditions were chosen empirically so that at a particular combination of temperature, back pressure, and flow rate--with that combination of settings defining solvent power, or more correctly, a solvent power distribution across the column, there was no peak distortion. Raising the temperature by small increments of 5 to 10 °C could produce peak distortion--presumably due to the decrease in solvent power by a decrease in density.** Increasing the back pressure slightly or increasing the flow rate (for an increased midpoint density) eliminated the

---

* $\rho_{out} \sim 0.35$ g/mL, 60°C, and FLOW setting = 2.00 mL/min--giving $\rho_{mid} \sim 0.40$ g/mL--resulted in distortion of anthracene peak on Hypersil SAS vs those values cited for distortion on the Hypersil SIL.

** $\rho_{mid} = 0.53$ g/mL--corresponding to 60°C, a back pressure of 1540 psig and a FLOW setting of 5.00 mL/min--gave no distortion for anthracene; however, $\rho_{mid} = 0.43$ given by 70°C and the same back pressure and FLOW setting gave distortion for anthracene on the fully active Hypersil SIL.

distortion at the higher temperatures. This described behavior seems to the author to be more related to some fundamental interaction of the solute, the chromatographic column and the mobile phase than to an injection procedure.

The chemical description of this interaction is still to be determined. It appears that there exists some threshold solvent power (defined either by the pure carbon dioxide density or the modifier identity and concentration in a modifier/carbon dioxide mixture) at which the solvent can begin to compete successfully with a particular stationary phase for a particular solute. Whether this involves a deactivation of active sites amenable to specific solute adsorption on the silica surface or a secondary solvent effect (43) where the mobile phase interacts with the solute as well as with the adsorption surface is unknown.

Conclusions and Future Experiments

SFC is competitive with other chromatographies in terms of chromatographic efficiency and resolution. The packed column SFC results should be applicable to capillary column SFC. The results of these preliminary experiments in establishing a modifier selection framework are gratifying in that dramatic differences in the chromatographic behavior were seen from vertex to vertex on the various modifier selectivity triangles. However, there are many studies to be undertaken to make the transition from preliminary results to a coherent, useful framework for routine modifier selection:

1. further preliminary surveys on various columns,

2. substitution of other modifiers for methylene chloride and chloroform if the preliminary results for other column stationary phases indicate a continued lack of suitability,

3. a statistical survey using a set of test solutes to yield a mathematical relationship that evaluates the importance of each of the available parameters--density, temperature, mobile phase selectivity and composition--for normal phase adsorption, normal phase partition, and reversed phase partition stationary phases in packed column studies and also for selected stationary phases in capillary column SFC,

   [At this point the density parameter appears to be less important in adsorption systems than the mobile phase composition; this is not true in reversed phase systems where density is quite important.]

4. continued evaluation of normal phase SFC compared to normal phase LC,

5. an in-depth study to determine the overall polarity of carbon dioxide and its selectivity fractions,

and 6. evaluation of "modifying the modifiers" by adding small quantities of strongly acidic or basic components, buffers, ion-pairing agents, and optically active agents to the major modifier component.*

Carbon dioxide based chromatography, with a mobile phase of carbon dioxide and mixtures of carbon dioxide plus modifiers ranging from subcritical to supercritical states, is a young but exciting area for study. Investigators can choose from a wide range of possible experiments: from fundamental physicochemical studies to semi-empirical frameworks to analytical methods-development for complex mixture separation. It is hoped that the preliminary results described in this paper will be useful to others in this area and will stimulate future work that will result in routine optimization schemes for SFC separations like those presently under development for LC separations.

EXPERIMENTAL

The conditions for the experiments described in this paper were the following: Sample Aliquot--0.5 µL of mixtures of approximately 1 mg/mL concentrations; Sample Take-Up Solvent--2-methoxyethanol, cyclopentane; Temperature--60 °C; Back Pressure--346 bar (5000 psig) unless specifically noted; Column--5 µm Hypersil SIL, 10 cm long, 4.6 mm ID, Hewlett-Packard #79916 SI Opt.554 for all presented chromatograms, (5 µm Hypersil SAS, 10 cm x 4.6 mm ID; material from Shandon Southern Products Limited, Cheshire, UK; packed in-house, and 5 µm Hypersil ODS, 10 cm X 4.6 mm ID, HP #79916 OD Opt. 554 for special discussions); Flow Setting--4 mL/min; Linear Velocity--0.5 cm/s; Detector Wavelengths--254 nm except for hormone detection at 240 nm; Mobile Phase--Pure CO_2, CO_2 plus modifiers; Modifier Concentration-10% (molar) maximum, premixed in commercial gas cylinder supplied to pump "B" and <10%, mixed with pure CO_2 supplied to pump "A"; and Instrument--Hewlett-Packard 1082 LC/SFC with added valve manifold.

---

*In the work described here, all concentrations of modifier in carbon dioxide were prepared by mixing pure carbon dioxide with a commercially prepared mixture of modifier and carbon dioxide (that mixture being in a standard gas cylinder) as outlined in the experimental section. Another way to prepare a modifier/carbon dioxide mixture is to pump pure liquid modifier with one pump and carbon dioxide with the other, mixing them in the instrument mixing chamber. Because of the large differences in fluid compressibilities, this method requires calibration to make a direct comparison to the first method (44). However, as an instrumental operating parameter, it is reproducible. Advantages are fast, easy modifier surveys and the possibility of "modifying the modifier," where a second modifier (possibly at very low concentrations and possibly more exotic than the major component) is combined with the first.

Sample

Seven different mixtures were prepared with a variety of test solutes; they are listed in Table II. These test solutes were obtained from a variety of sources in a broad range of purities (44). Take-up solvents, 2-methoxyethanol and cyclopentane, were spectrograde HPLC solvents (Burdick and Jackson Laboratories, Inc.).

Mobile Phase

The mobile phase was carbon dioxide and carbon dioxide plus modifier where the modifiers were 2-propanol, 2-methoxyethanol, chloroform, and methylene chloride. The carbon dioxide/modifier mixtures were purchased in conventional gas cylinders with dip tubes from Scientific Gas Products, Inc. (South Plainfield, New Jersey) who prepared them as requested as Class C mixtures (Certified Standard) with Instrument Grade carbon dioxide (99.99% minimum purity) and HPLC spectro-grade liquid solvents (Burdick and Jackson Laboratories, Inc.). The pure carbon dioxide was also Instrument Grade carbon dioxide supplied in cylinders with dip tubes by Scientific Gas Products, Inc.

A valve manifold, Figure 13, was added to the standard Hewlett-Packard 1082 HPLC/SFC configuration so that any four of the possible five solvent gases could be valved to either of the pumpheads. This ability to combine modifier/carbon dioxide solutions was provided in anticipation of conducting a detailed statistical experiment where the vertices, the midpoints of the sides and the center point of the modifier selectivity triangle will be studied in the prescribed random fashion. In the experiments described here, pure carbon dioxide was was pumped by "A" pump and the carbon dioxide/modifier mixtures were pumped by "B" pump. The maximum modifier concentration in the mobile phase was 10% (molar), corresponding to "100% B," an instrument operating parameter. Lower modifier concentrations were obtained by operating the instrument between 15% B and 95% B--i.e., pure carbon dioxide was mixed with the maximum concentration carbon dioxide/modifier mixture. While it is certainly possible to do gradient programming to change the modifier concentration in the mobile phase continuously, these experiments were conducted isocratically. With this particular kind of normal phase column, it was found that the desired modifier concentration in equilibrium with the column was quickly achieved when that parameter was ballistically changed, taking only 2-5 minutes as as indicated by retention time stabilities. The experimental procedure providing the evidence on which the last statement is based is as follows: 1. order a ballistic change in %B, 2. note the time as soon as the instrument flow control light indicates the flow is stable (1-3 minutes, depending upon the size of the change in %B), 3. start injecting samples any time after "FLOW STABLE" until the component retention times are stable, and 4. periodically check for stable retention times at very long times after "FLOW STABLE." Plotting the detector signal throughout the concentration change shows a distinct and rapid step in the signal baseline as the change in concentration takes place.

Chemical traps (CT) as described in Figure 13 were placed immediately downstream of the solvent cylinders. Their function was to remove the trace amount of water remaining in Instrument Grade carbon dioxide as well as any other trace contaminants present. Studies are presently underway to establish the need and determine the composition of the chemical traps.

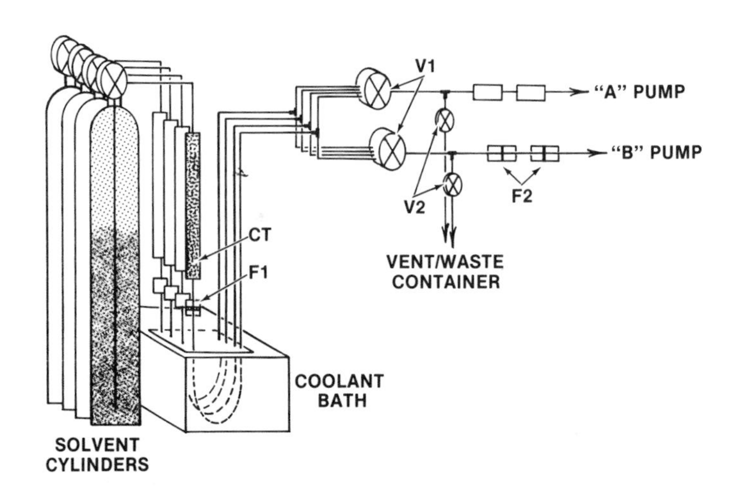

Figure 13. Solvent supply system showing selection valve manifold. All components downstream of the coolant bath to pump inlets are thermally insulated.

CT - chemical traps: 120-150 mL of Silica Gel 60 (230/400 mesh, EM Reagents, methanol-rinsed, activated at 110 °C, 20 hours) in high pressure traps (prototypes, Adsorbents and Dessicants of America)
F1 - filter units supplied in SFC modification kit (sintered stainless steel frits: 2 µm, 1/32" x 0.5" OD); HP#79887-60620
V1 - stream selection, 5-way ball valves (Whitey Co.; Part #SS43Z)
V2 - On-off ball valve (Whitey Co.; Part #SS41S2)
F2 - Filters: made from zero volume unions with 1/8" OD X 2 µm screens. (Both components, Valco Instruments Co., Inc.; union--part #ZVU-125) Union ID enlarged to 0.073" from 0.029"; counterbore machined in tubing ends against screens
Solvent cylinders - see text
Coolant bath - Neslab, Model RTEZ-4 (-30 to 100 °C, 5 L, 765 BTU/hr at -10°C, 0-13 L/min at 0 head, ethylene glycol/ water solution)

Temperature

A standard operating temperature of 60 C was chosen for two reasons. First, a rough approximation for estimating binary mixture critical points is that (15)

$$T_{cm} = X_a T_{ca} + X_b T_{cb}$$

where T_{cm} is the mixture critical temperature, T_{ca} and T_{cb} are the component critical temperatures, and X_a and X_b are the component mole fractions. For all of the maximum 10% (molar) mixture concentrations, mixture critical temperatures estimated in this way lie between 50-55°C, so a temperature of 60 °C was chosen so that the mobile phase was ostensibly a supercritical mixture. Second, the modest value of 60°C was chosen versus some other value up to the maximum of 100 °C possible with this instrument to emphasize the applicability of SFC to separations of thermally labile compounds.

Back Pressure

Density is an important "extra" variable that is available in SFC and not in GC and HPLC. However, in this work the goal was to study the effect of the modifier specifically, so this parameter was held near its maximum. For pure carbon dioxide, an outlet pressure of 346 bar at 60°C corresponds to a column outlet density of about 0.86 g/mL; column inlet pressures with the noted flow rates were usually about 375 bar (5450 psig), corresponding to a column inlet density of 0.88 g/mL for pure carbon dioxide.

Column

A secondary goal of these preliminary studies was to characterize supercritical fluid normal phase adsorption chromatography: if column re-equilibration times were much shorter than those encountered with typical liquid mobile phases in adsorption chromatography, a revival of normal phase adsorption chromatography with all of its advantages would be possible. In particular, this chromatographic method is suitable for separations according to chemical type and to differing numbers of functional groups and especially for separations of isomers of organic-soluble, intermediate molecular weight (100 to 200 g/mole), nonionic solutes (45). Therefore, a Hypersil SIL column was chosen as the first column. Because of the chemical traps upstream of the pumps for dry, purified carbon dioxide and carbon dioxide/modifier mixtures along with column activation at 100°C with the pure, dry carbon dioxide, the analytical silica adsorbent would have been essentially fully active. This is different from conventional liquid-solid chromatography where there is usually some chosen degree of adsorbent deactivation by water or an alcohol.

Literature Cited

1. T.H. Gouw, R.E. Jentoft, and E.J. Gallegos, High Pressure Science and Technology, Sixth AIRAPT Conference, (K.D. Timmerhaus and M.S. Barber, ed's.), Plenum Press, New York, 1979, Vol. 1, p. 583.
2. T.A. Rooney, High Resolution Gas Chromatography, (R.R. Freeman, ed.), Hewlett-Packard (#5950-3562), Avondale, Pennsylvania, 1979, Chapt. 1.
3. W. Jennings, Gas Chromatography with Glass Capillary Columns, Academic Press, New York, 1980, Chapters 1 and 6.
4. L.R. Snyder and J.J. Kirkland, Introduction to Modern Liquid Chromatography, Second Edition, John Wiley & Sons, Inc., New York, 1979, Chapt. 5.
5. Unpublished data, courtesy of T.N. Tweeten, Hewlett-Packard, Avondale, Pennsylvania.
6. J.C. Giddings, Dynamics of Chromatography, Marcel Dekker, New York, 1965.
7. D.R. Gere, Science (October 31, 1983), in press.
8. C.P.M. Schutjes, E.A. Vermeer, J.A. Rijks, and C.A. Cramers, Proceedings of the Fourth International Symposium on Capillary Chromatography, (R.E. Kaiser, ed.), Institute of Chromatography, Bad Durkheim, 1981, p. 687.
9. C.A. Cramers, F.A. Wijnheymer, and J.A. Rijks, J. High Resolut. Chromatogr. Chromatogr. Commun. 2, 329 (1979).
10. G.M. Schneider, Angew. Chem. Int. Ed. Engl. 17, 716 (1978).
11. L.M. Bowman, Ph.D. Thesis, University of Utah (1976).
12. D.R. Gere, R. Board, and D. McManigill, Anal. Chem. 54, 736 (1982).
13. R. Feist and G.M. Schneider, Sep. Sci. Technol. 17(1), 261 (1982).
14. P.A. Peaden, J.C. Fjeldsted, M.L. Lee, S.R. Springston, and M. Novotny, Anal. Chem. 54, 1090 (1982).
15. R.B. Bird, W.E. Stewart, and E.N. Lightfoot, Transport Phenomena, John Wiley & Sons, Inc., New York, 1960.
16. L.G. Randall, Sep. Sci. Technol. 17(1), 1 (1982).
17. H.E. Weaver and P.F. Bente, "Solvatochromic Effects of Mesityl-oxide and Pyrazine in Fluid Carbon Dioxide," Presented at the Fall Meeting of the American Chemical Society/Physical Chemistry Division, Washington D.C., 1983, Paper 23.
18. P.F. Bente and H.E. Weaver, "Hildebrand Solubility Parameters for Liquid and Supercritical Carbon Dioxide," Presented at the Fall Meeting of the American Chemical Society/Physical Chemistry Division, Washington, D.C., 1983, Paper 24.
19. L.M. Bowman, M.N. Myers, and J.C. Giddings, Sep. Sci. Technol. 17(1), 271 (1982).
20. U. van Wasen, I. Swaid, and G.M. Schneider, Angew. Chem. Int. Ed. Engl. 19, 575 (1980).
21. S.T. Sie and G.W.A. Rijnders, Sep. Sci. 2(6), 729 (1967).

22. E.N. Fuller, P.D. Schettler, and J.C. Giddings, Ind. Eng. Chem. 58(5), 19 (1966).
23. C.P.M. Schutjes, E.A. Vermeer, J.A. Rijks, and C.A. Cramers, J. Chromatogr. 253, 1 (1982).
24. J.C. Giddings, M.N. Myers, L. McLaren, and R.A. Keller, Science 162, 67 (1968).
25. A.W. Francis, J. Phys. Chem. 58, 1099 (1954).
26. T.H. Gouw, J. Chem. Eng. Data 14(4), 473 (1969).
27. G. Brunner, Habilitationsschrift, Universitat Erlangen-Nurnberg, Erlangen (1978).
28. R.A. Snedeker, Ph.D. Dissertation, Princeton University (1955).
29. L.R. Snyder, J. Chromatogr. 92, 223 (1974).
30. L.R. Snyder, J. Chromatogr. Sci. 16, 223 (1978).
31. J.L. Glajch, J.J. Kirkland, and L.R. Snyder, J. Chromatogr. 238, 269 (1982).
32. L.R. Snyder and J.L. Glajch, J. Chromatogr. 248, 165 (1982).
33. L.R. Snyder, J.L. Glajch, and J.J. Kirkland, J. Chromatogr. 218, 299 (1981).
34. L.R. Snyder and J.L. Glajch, J. Chromatogr. 214, 1 (1981).
35. J.L. Glajch and L.R. Snyder, J. Chromatogr. 214, 21 (1981).
36. J.L. Glajch and J.J. Kirkland, Anal. Chem. 54 (14), 2593 (1982).
37. J.L. Glajch, J.J. Kirkland, K.M. Squire, and J.M. Minor, J. Chromatogr. 199, 57 (1980).
38. L.R. Snyder, Chemtech, 750 (December 1979).
39. L.R. Snyder, Chemtech, 188 (March 1980).
40. J.H. Hildebrand and R.L. Scott, Regular Solutions, Prentice-Hall, Inc., Englewood Cliffs, New Jersey, 1962.
41. L. Rohrschneider, Anal. Chem. 45(7), 1241 (1973).
42. E. Lederer and M. Lederer, Chromatography: A Review of Principles and Applications, Elsevier Publishing Company, New York, 1957, Chapt. 5.
43. L.R. Snyder, Principles of Adsorption Chromatography, Marcel Dekker, Inc., New York, 1968.
44. L.G. Randall, Technical Paper No. 102, Hewlett-Packard, Avondale, Pennsylvania (1983).
45. L.R. Snyder and J.J. Kirkland, Introduction to Modern Liquid Chromatography, John Wiley & Sons, Inc., New York, 1974, Chapt. 8.
46. J.J. Kirkland, J. Chromatogr. 83, 149 (1973).

RECEIVED November 10, 1983

12

Super Resolution in Chromatography by Numerical Deconvolution

JAMES B. CALLIS

Department of Chemistry, BG-10, University of Washington, Seattle, WA 98195

Super-resolution in chromatography is defined as the ability to detect the presence of two or more components when their mean elution times fall within two standard peak widths of each other. Numerical deconvolution techniques for achieving super-resolution are considered for the case of a single channel detector (FID), a multichannel detector (MS) and a two dimensional detector (video fluorometer). It is shown that the degree of ambiguity of the deconvolution decreases in going from a monochannel detector to a multichannel detector to a multidimensional detector. Three types of analytical situations are considered for each type of detector: (a) a quantitative analysis is performed where all of the components are knowns for which calibration data is available, (b) a qualitative analysis is performed obtaining the retention times and spectra of each component, and (c) a mixed analysis is performed quantitating the known components and qualitating the unknowns. It is shown that it is possible to achieve super-resolution in all cases, but the quality of the results increases as the dimensionality of the data set increases.

As the proceedings of this symposium attest, major technological advances in high resolution chromatography have recently been made. For example, 30 meter capillary gas chromatography columns are commercially available which routinely provide resolving powers of over 100,000 theoretical plates, and additionally yield near-ideal (gaussian) peak shapes for a wide variety of compounds of disparate polarity. In liquid chromatography, similar advances in technology are also beginning to appear and become commercially available. As wonderful as these achievements seem, there still remains an urgent need for even higher resolution; especially if we are ever to solve the

0097-6156/84/0250-0171$07.75/0

problem of a global analysis of an extremely complex mixture such as a combustion sample or drugs in human body fluids. Recently, three very sobering papers have appeared which provide a basis for estimating the occurrence of component overlap based upon a random distribution of retention times (1-3). For example, Davis and Giddings (3) have shown that a column which is capable of resolving 100 components, if they elute sequentially at equal time intervals, can resolve only 18 of 50 components into single peaks in the case where they elute at random intervals. Alternatively, these authors (3) demonstrate convincingly, "that a chromatogram must be approximately 95% vacant in order to provide a 90% probability that a given component of interest will appear as an isolated peak." While it is certain that technological advances will continue to provide greater numbers of theoretical plates, it is also certain that the combination of the fundamental limitations of the physics of the separation process, together with the accelerating demands for analysis of ever more complex substances, will combine to assure that we will never have enough theoretical plates. Our purpose in this paper is to show that additional resolution can be obtained by numerical techniques. In particular, we shall concentrate on algorithms which can detect and quantitate the presence of two or more components when their mean retention times lie within two standard peak widths (standard deviations) of each other. As Figure 1 shows, when this condition occurs, it is virtually impossible to detect the presence of multiple components by eye. We call the ability of numerical techniques to resolve such fused peaks "super-resolution" because we have, in effect, increased the resolving power of the chromatogram beyond that implied by the column's number of theoretical plates.

We begin our exposition by first considering one dimensional chromatograms where the output of the column is monitored by some sort of linear single channel detector (FID, thermocouple, index of refraction, etc.). An algorithm will be developed which can achieve super-resolution, as well as provide an estimate for the number of components contributing to a given peak. We will then consider the case where there is a multichannel detector, each channel of which has a different response to each different component of the mixture being separated. In this case, the measured output is a function of two variables - retention time and detector channel. For such two dimensional data sets, more powerful deconvolution techniques such as factor analysis and rank-annihilation can be used. Finally we will consider the analysis of higher dimensionality chromatographic data, as might be obtained using a video fluorometer or ms-ms detector. Here, the data can be represented as a tensor of rank three. The deconvolution of such data must be solved by computationally expensive iterative methods, but the rewards are great, for it will be shown that if a solution can be found, it is the correct one.

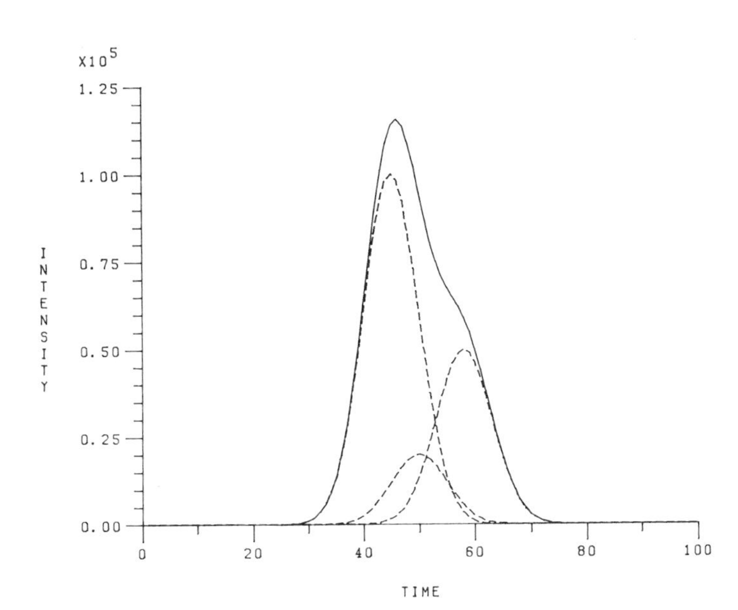

Figure 1. A computer simulated example of the need for super-resolution in chromatography. The composite curve, shown as a solid line, is the sum of three gaussian components occurring at time channels 45, 50 and 58, respectively with peak maxima ratios of 1:2:5. All peaks have a standard peak width of 5. The individual gaussians are shown as dashed lines.

Monodetector Chromatography

In this section we will consider a simple chromatogram generated by some type of single channel detector (FID, EC, thermocouple, etc.). We will confine our treatment to the case where the response of the detector is linear with mass and the elution profile of a component is independent of its amount and the presence of all other components. If these conditions hold, the chromatogram can be expressed in the following form:

$$\bar{I}(t) = \sum_{k=1}^{p} \alpha^k \bar{r}^k(t) \qquad [1]$$

where $\bar{I}(t)$ is a vector quantity giving the observed response of the detector as a function of time after injection, $\bar{r}^k(t)$ is the normalized retention vector for the kth component, and α^k is a scalar calibration factor relating the observed detector response to the quantity of the kth component injected. Equation 1 is a simple, linear model of the data. In the most general case, the analysis of Equation 1 consists of finding p, and the α^k and $\bar{r}^k(t)$ for each of the p components. Because of the difficulty of such an analysis in the case where overlap of components is severe, let us consider first the more tractable situation where the p components are all knowns for which standard calibration runs are available, giving the p $\bar{r}^k(t)$'s and α^k's at specific concentrations. Under these circumstances, the analysis of an unknown mixture consists of finding the α^k's, which when divided by the standard α^k's gives the quantity of each of the k components in the mixture. As is well known (4), the best approach to solving Equation 1 for the α^k's is by the method of least squares or, in the case where error estimates for the $\bar{r}^k$'s are available, by the maximum likelihood estimate. In the latter case, if we suppose that we have a data set $\{I_i\}$ consisting of n measurements of $\bar{I}(t)$ at equal time intervals indexed by i, as well as data sets for each $\bar{r}^k = \{r_i^k\}$ obtained in separate experiments, then we can write an expression for χ^2, the statistically weighted deviations of the model (Equation 1) from the data

$$\chi^2 = \sum_i \left\{ \frac{1}{\sigma_i^2} \left[I_i - \sum_k \alpha^k r_i^k \right]^2 \right\} \qquad [2]$$

where σ_i is the root mean square error in the intensity measurement at the ith interval. The "maximum likelihood" estimates for α^k are found when

$$\frac{\partial \chi^2}{\partial \alpha^k} = 0 \qquad [3]$$

is satisfied for each α^k. The solution of these equations can be expressed as

$$\sum_i \left[\frac{1}{\sigma_i^2} I_i r_i^j\right] = \sum_k \left\{\alpha^k \sum_i \left[\frac{1}{\sigma_i^2} r_i^k r_i^j\right]\right\} \quad [4]$$

Figure 2 shows the use of the linear least squares technique to deconvolute the computer synthesized composite chromatogram of Figure 1 into its three components. In Figure 2, the input data is represented as the points, while the "best fit" is represented as the solid line. Even though the middle peak is buried in the valley between the two major peaks and is within one standard deviation of the left-most peak as well, it is readily quantitated. Beneath this curve is plotted the residuals, which are simply the difference between the input intensities and those predicted by the model at each time interval. In this case, the residuals vary randomly from point to point and the size of the variation is consistent with the amount of simulated noise added to the data. Other examples of least squares fitting of overlapping peaks to chromatographic (5) and spectroscopic (6-9) data can be found in the literature. Thus, super-resolution is easily achieved in the case where all of the components are knowns for which standard runs are available.

Of course, one is seldom blessed with samples which are free from unknown peaks which overlap the known peaks and interfere with their determination. Therefore, it is desirable to have techniques which can quantitate a subset of known components in a complex, unresolved mixture. Clearly, an unconstrained least squares procedure as given by Equation 4 will not suffice, as shown in Figure 3 where the data of Figure 1 is now fit to two of the three components. While the data appears fairly well modeled by two gaussian peaks, examination of the residuals shows that there are systematic deviations between the data and the model which are larger than the noise. Obviously, other interpretations of the data exist. The importance of examining the residuals for systematic deviations between model and data has been previously discussed by Grinvald and Steinberg (10) and cannot be overemphasized.

It is possible to arrive at a better estimate for the known components using a constrained least squares approach. In the case where the data can be modeled by Equation 1, and only one component is a known, the following relationship is certainly valid:

$$\bar{I}(t) - \alpha^{1}\bar{r}^{1}(t) \geqslant 0 \quad \text{for all } t \quad [5]$$

or

$$0 \leqslant \alpha^1 \leqslant \min_t \left[\bar{I}(t)/\bar{r}^1(t)\right] \quad [6]$$

If there is any time at which this molecular species is the only

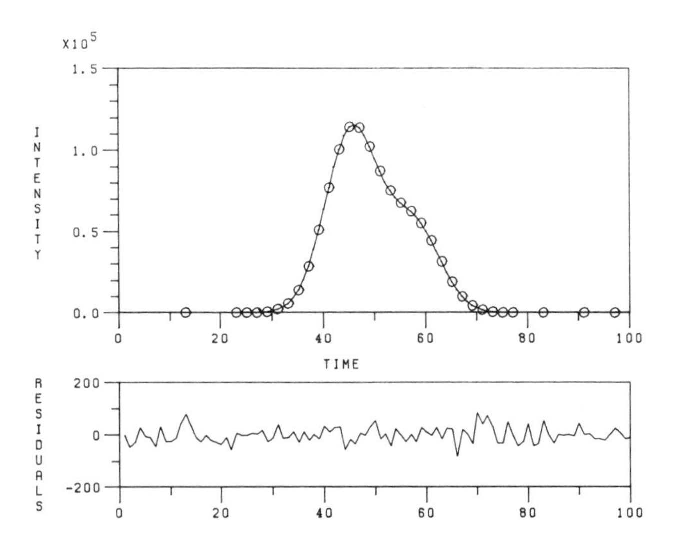

Figure 2. Least squares analysis via Equation 4 of the data of Figure 1, represented as the circles. It was assumed that the peak positions and standard peak widths of all three components were known, leaving only the relative contribution of each to be determined. Random noise was added to the data to simulate a signal-to-noise ratio of 1000/1. The "best fit" is shown as a solid line. The lower figure is a plot of the residuals.

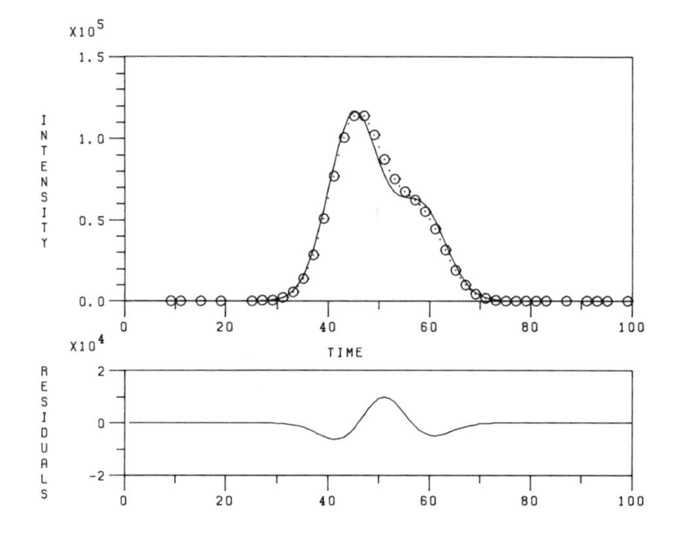

Figure 3. Reanalysis of the data of Figure 1 (with noise), assuming only two components are known (the ones centered at 45 and 58).

component eluting, then $\min_t\{\bar{I}(t)/\bar{r}^{1}(t)\}$ will actually equal α^1. When there are s known components present (s<r), we have, in analogy with Equation 2,

$$\bar{I}(t) - \sum_{k=1}^{s} \alpha^k \bar{r}^k(t) \geqslant 0 \qquad [7]$$

This equation forms the basis for the use of a "non-negative least sum of errors" approach to finding the s values of α^k for which calibration data exists (11). Essentially, we find α^k subject to minimizing χ^2 as per Equation 2 and subject to the i inequality constraints as expressed by Equation 7. Unfortunately, this approach is expected to work well only in the case where each of the known components is, individually, the only one eluting. In the case of severe overlap of components, there will be many solutions which can satisfy the constraints and thus the results will be ambiguous. The problem with partial analysis of data of the form of Equation 1 is that we do not know the value of p; without this information, the relative contributions of the knowns and unknowns cannot in general be unambiguously assigned. However, if we can make one more assumption about the form of the data, we can then show that the range of possible solutions consistent with the data is so severely constrained, that in favorable circumstances, p can be accurately determined.

In order to proceed, we will need to introduce two hypothetical constructs. The first of these is the infinite resolution or perfect chromatogram, $\bar{I}^{per}(t)$, which has the following form:

$$\bar{I}^{per}(t) = \sum_{k=1}^{p} \alpha^k \delta^k\left[t-t^k\right] \qquad [8]$$

where $\delta^k(t-t^k)$ is the dirac delta function and t^k is the retention time of the kth component. Here, obviously, there is a finite number of components and all of them are resolved. Of course no real chromatogram ever obeys Equation 8, and one must therefore include some type of peak shape function, g(t,t'). In the case where g(t,t') is (a) continuous, finite valued and differentiable, and (b) a function of only retention time and no component dependent variable, we can define the ideal chromatogram, $\bar{I}^{id}(t)$, as follows:

$$\bar{I}^{id}(t) = \int_{-\infty}^{\infty} \bar{I}^{per}(t')g(t,t')dt' \qquad [9]$$

In this approximation, every component in the chromatogram is subject to the same peak broadening mechanism, and therefore we can hope to obtain g(t,t') experimentally by chromatographing standard components whose retention times are in the range covered by the unknown mixture. In the case of capillary gas chromatography, we can reasonably expect the peak shapes to be gaussian. For an isothermal run, the mean retention time divided by the standard deviation will be a constant, while for a temperature programmed run, the peak parameters will be independent of retention time, and thus $g(t,t') = g(t-t')$. In this case Equation 9 becomes

$$\bar{I}^{id}(t) = \int_{-\infty}^{\infty} \bar{I}^{per}(t')g(t-t')dt \qquad [10]$$

Thus equation 10 represents the <u>convolution</u> of the peak broadening function with the perfect chromatogram, and is widely known as the convolution equation.

We now see that our problem of determining the p components may be approached by solving Equation 9 or 10 for $\bar{I}^{per}(t)$, given $\bar{I}^{id}(t)$ and $g(t,t')$. Equations of this form are widely used to describe physical phenomena of various types. A very lucid description of possible approaches to solving Equations 9 and 10 when they are applied to optical image enhancement problems is given by Frieden (<u>12</u>). This author shows that although Equations 9 and 10 are linear, their solution by standard linear methods will generally be unsatisfactory. For example, one very straightforward approach is to write Equation 9 in discrete form.

$$I_i^{id} = \sum_{j=1}^{n} I_j^{per} g_{ij} \quad ; \quad \bar{I}^{id} = \bar{I}^{per}\underset{\sim}{G} \qquad [11]$$

where $\bar{I}^{id}(t) \equiv \{I_i^{id}\}$ is the vector representing the measured chromatogram, $\bar{I}^{per}(t) \equiv \{I_j^{per}\}$ is the vector representing the (sought-for) perfect chromatogram, and $\underset{\sim}{G} = \{g_{ij}\}$ is a matrix which represents the peak broadening mechanism. In the case where i and j have the same dimension, we can solve Equation 11 as

$$\bar{I}^{per} = \bar{I}^{id}\underset{\sim}{G}^{-1} \qquad [12]$$

Unfortunately, this solution is seldom satisfactory because (a) it does not take into account the presence of noise in a satisfactory way, which often leads to idiopathic oscillations, and (b) it does not constrain the solution to be physically reasonable (e.g. all positive). As Freiden has observed, there are methods which can overcome the limitations of linear solutions, but these are non-linear and thus are more difficult to carry out. Nevertheless, the rewards are great as will be seen.

Let us now introduce two constraints on the solution which will greatly decrease the degree of ambiguity. The first is that the components of the mixture are eluted over some finite time interval, i.e. that $\bar{I}^{per}(t)$ is known to be zero outside this interval. When this condition holds we immediately see that there are less parameters to estimate ($\{I_j^{per}\}$) than data points which define them. Thus Equation 11 can be solved by the method of least squares, using essentially the same set of equations as Equation 4.

Unfortunately, the use of the linear least squares approach will not always yield physically reasonable estimates of $\bar{I}^{per}(t)$. This is because there will likely be solutions which will give very good values of χ^2 but for which $\bar{I}^{per}(t)$ goes negative. Thus we need to constrain the solutions to those where $\bar{I}^{per}(t) \geqslant 0$. Fortunately, the problem of linear least squares with linear inequality constraints has been considered in detail by Lawson and Hanson (13). In Figure 4 we show how the data of Figure 1 can be analyzed using the non-negative least squares approach. The original input data, consisting of the simulation of three overlapping components (Figure 1), is redisplayed in Figure 4 as the circled points. The estimated solution for $\bar{I}^{id}$ is displayed as the solid line. Clearly, it consists of three delta functions, each corresponding in both intensity and position to the input components. In Figure 4, we have also displayed the residuals, which are the differences between the input data and the recovered ideal chromatogram convoluted with the peak broadening function. The residuals vary randomly as expected. As can be seen, the algorithm correctly determines the number of components, and estimates their relative peak heights as well. Thus, we now have a reasonable method for estimating the number of components in an ideal chromatogram even when the peaks overlap so severely their presence is not detectable by eye. The major limitation of this algorithm is that it requires a very high S/N (at least 100/1). Fortunately, there are detectors which can provide such data. Further work is in progress to determine how well this algorithm performs on a variety of real data, including the situation where some of the peak shapes are non-ideal.

Multidetector or Multichannel Detector Chromatography

In multidetector chromatography, the effluents of the chromatogram are monitored by multiple detectors (e.g., FID, EC and N/P detectors) or by a multichannel spectrometric detector (UV-VIS absorbance, IR, MS). Each channel or detector is assumed to have a different linear response to each component in the mixture. In this case our chromatogram is now two dimensional; i.e., the observed intensity is a function of two variables: (a) the chromatographic retention time, t, and (b) the detector channel number, λ. Thus, we shall write the observed intensity as $\tilde{M}(t,\lambda)$ to distinguish it from the one dimensional case represented by $\bar{I}(t)$. Under the following assumptions the data set from a multidetector chromatography run has a linear form: (a) the response of each detector is linearly related to the amount of each component present, and (b) the response to a particular component is independent of all other components. For

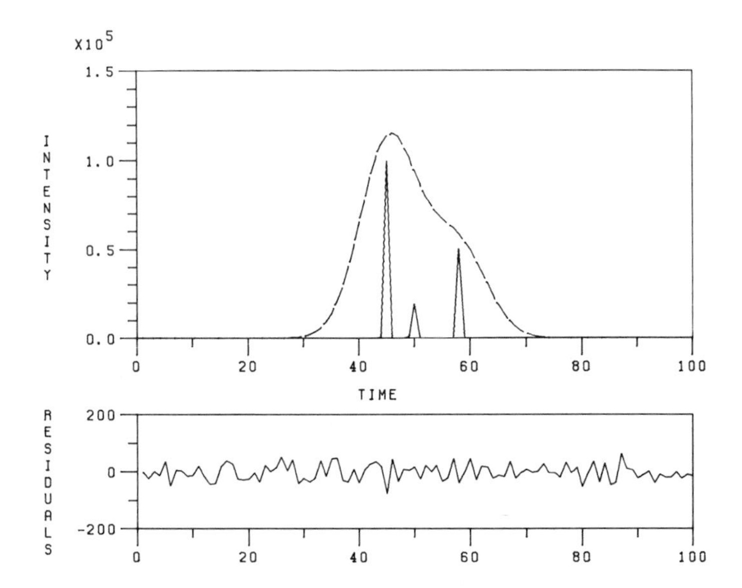

Figure 4. Reanalysis of the data of Figure 1 (with noise) via the method of non-negative least squares, making no assumptions about the number of knowns.

a mixture of p components the data set can be expressed as a matrix, $\underset{\sim}{M}$, which has the following form:

$$\underset{\sim}{M} \equiv (m_{ij}) = \left\{ \sum_{k=1}^{p} \alpha^k r_i^k s_j^k \right\} \quad ; \quad \underset{\sim}{M} = \sum_{k=1}^{p} \alpha^k \bar{r}^k * \bar{s}^k \qquad [13]$$

where r_i^k is the fraction of the kth component which elutes at the ith time interval. The time sequenced set $\{r_i^k\}$ may be thought of as the normalized retention vector for the kth component, as before. s_j^k is the normalized response of jth detector or detector channel to the kth component, α^k is a scalar which contains all of the concentration dependence of the kth component's contribution to $\underset{\sim}{M}$, and the symbol * denotes the outer product operation. In the case where $\{s_j^k\}$ represents the output of a multichannel detector, such as a mass spectrometer, $\bar{s}^k \equiv \{s_j^k\}$ can be thought of as the spectral vector for the kth component.

According to Equation 13, the most general analysis of $\underset{\sim}{M}$ consists of finding p, the number of components, and the α^k, $\bar{r}^k$ and $\bar{s}^k$ for each component. In many situations, the p components are knowns for which the $\bar{r}^k$'s and $\bar{s}^k$'s have been previously determined, thus all that is needed are the α^k's. In this situation one can use the technique of least squares. Equation 13 may be expressed as

$$\underset{\sim}{M} = \sum_{k=1}^{p} c^k \underset{\sim}{M}^k \qquad [14]$$

where the $\underset{\sim}{M}^k$'s are the standard matrices for each of the components, and the c^k's are sought-for concentrations divided by the standard concentrations. Equation 14 is of the same form as Equation 2, and the best fit may be found by analogy with Equations 2-4 from the following set of equations:

$$\sum_k \left[\sum_{i,j} (m_{ij})^k (m_{ij})^l \right] \alpha^k = \sum_{i,j} (m_{ij})^l m_{ij} \qquad [15]$$

where m_{ij} is a matrix element of the observed data matrix, and $(m_{ij})^k$ and $(m_{ij})^l$ are the matrix elements of the standard data matrices for the kth and lth components.

Let us now consider the analytical situation where the mixture is totally unknown. In this case, when the number of components is small, it is possible to find p and estimates for each of the $\bar{r}^k$'s and $\bar{s}^k$'s. Essentially, we use the techniques of factor analysis (14) to deconvolute the data. The first task is to find an estimate for p. It can be readily shown that the rank of $\underset{\sim}{M}$ is a lower bound to the number of linearly independent components. The best procedure for finding the rank of a matrix $\underset{\sim}{M}$ is to construct the eigenvalues $\{\xi_k\}$ and eigenvectors $\{\bar{u}^k, \bar{v}^k\}$ of the real, square symmetric matrices $\underset{\sim}{M}\underset{\sim}{M}^T$ and $\underset{\sim}{M}^T\underset{\sim}{M}$, where $\underset{\sim}{M}^T$ is the transpose of M. These are obtained from the following equations:

$$\bar{v}^{k}\underset{\sim}{M}^{T}\underset{\sim}{M} = \xi_k \bar{v}^{k} \quad ; \quad \underset{\sim}{M}\underset{\sim}{M}^{T}\bar{u}^{k} = \xi_k \bar{u}^{k} \qquad [16]$$

Ideally, the number of non-zero eigenvalues will equal the rank of the matrix. Unfortunately, because of experimental and computational errors, none of the eigenvalues are really zero. Thus one has the practical problem of deciding which are due to signal and which to noise. In favorable cases, there will be a few eigenvalues which are rather larger than the rest and the number of these will provide the correct estimate for p.

Eigenanalysis allows us to proceed even further and to obtain estimates for the p $\bar{r}^{k}$'s and $\bar{s}^{k}$'s. These are obtained from the sets of eigenvectors which correspond to the p chosen eigenvalues. It can be readily shown that the set of vectors $\underset{\sim}{U}=(\bar{u}^{1},\bar{u}^{2},\ldots,\bar{u}^{p})$ is congruent to the set $\underset{\sim}{R}=(\bar{r}^{1},\bar{r}^{2},\ldots,\bar{r}^{p})$ while the set $\underset{\sim}{V}=(\bar{v}^{1},\bar{v}^{2},\ldots,\bar{v}^{p})$ is congruent to $\underset{\sim}{S}=(\bar{s}^{1},\bar{s}^{2},\ldots,\bar{s}^{p})$. And even if we have not chosen the value of p correctly it is still true that the quantity $\sum^{p}\sqrt{\xi}^{k}\bar{u}^{k}\bar{v}^{k}$ provides the best approximation of rank p to M, in the least squares sense.

Unfortunately, finding the sets $\underset{\sim}{U}$ and $\underset{\sim}{V}$ does not quite solve the problem of approximating $\underset{\sim}{R}$ and $\underset{\sim}{S}$, because they will contain negative elements (to satisfy orthogonality conditions), while the sets of $\bar{r}$ and $\bar{s}$ vectors must be all-positive. However, it is readily shown that there exists a transformation matrix K which transforms the $\underset{\sim}{U}$ and $\underset{\sim}{V}$ matrices to matrices $\underset{\sim}{U}'>0$ and $\underset{\sim}{V}'>0$ by the following rules:

$$\underset{\sim}{U}' = \underset{\sim}{U}\underset{\sim}{K} \quad ; \quad \underset{\sim}{V}' = \underset{\sim}{V}\left(\underset{\sim}{K}^{T}\right)^{-1} \qquad [17]$$

Essentially, the matrix $\underset{\sim}{K}$ may be thought of as an "oblique" transformation of the data. We must emphasize here that the $\underset{\sim}{U}$ and $\underset{\sim}{V}$ matrices are not rotated independently; if $\underset{\sim}{K}$ is used to transform $\underset{\sim}{U}$, then $(\underset{\sim}{K}^{T})^{-1}$ must be used to transform $\underset{\sim}{V}$.

Because of (a) the overlaps of the spectral and retention time vectors, and (b) the noise in the data, there will in general be a range of possible $\underset{\sim}{K}$'s which can transform the $\underset{\sim}{U}$ and $\underset{\sim}{V}$ sets to sets with all-positive elements. Thus the estimates for the spectral vectors will have some degree of ambiguity in them. The exact degree of uncertainty has been shown to be dependent on the degree of overlap of the spectra and retention profiles and whether there are regions in the time, detector channel plane where only one component contributes to the observed intensity. By working backward from known spectra and retention time profiles with various overlaps it is possible to determine the degree of ambiguity in the deconvolution procedure for the two component case. In Table I we have summarized our findings (15) for the 15 conceivable types of overlap. (The case of total overlap is not considered because it will yield a rank

Table I. Prediction of the ambiguities inherent in the transformation of the eigen vectors of a two component matrix to all-positive spectral and retention vectors. The k's are the off diagonal elements of the transformation matrix K.

Retention Vector Overlap (columns); Spectral Vector Overlap (rows)

	1, 2	1, 2	1, 2	1, 2
 1, 2	$\bar{r}^{(1)}, \bar{r}^{(2)}, \bar{s}^{(1)}, \bar{s}^{(2)}$, uncertain Both k elements extreme	$\bar{r}^{(2)}, \bar{s}^{(1)}$, uncertain one k unique one k extreme	$\bar{r}^{(2)}, \bar{s}^{(1)}$, uncertain one k unique one k intermediate	$\bar{r}^{(1)}, \bar{r}^{(2)}, \bar{s}^{(1)}, \bar{s}^{(2)}$ uncertain one k extreme one k intermediate
 1, 2	$\bar{r}^{(1)}, \bar{s}^{(2)}$, uncertain one k unique one k extreme	no uncertainties Both k's unique	$\bar{r}^{(2)}, \bar{s}^{(1)}$, uncertain one k unique one k extreme	$\bar{r}^{(1)}, \bar{r}^{(2)}, \bar{s}^{(1)}, \bar{s}^{(2)}$ uncertain Both k's extreme
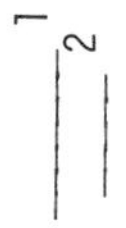 1, 2	$\bar{r}^{(1)}, \bar{s}^{(2)}$, uncertain one k unique one k intermediate	$\bar{r}^{(1)}, \bar{s}^{(2)}$, uncertain one k unique one k extreme	$\bar{r}^{(1)}, \bar{r}^{(2)}, \bar{s}^{(1)}, \bar{s}^{(2)}$ uncertain Both k elements extreme	$\bar{r}^{(1)}, \bar{r}^{(2)}, \bar{r}^{(1)}, \bar{r}^{(2)}$ uncertain one k extreme one k intermediate
 1, 2	$\bar{r}^{(1)}, \bar{r}^{(2)}, \bar{s}^{(1)}, \bar{s}^{(2)}$ uncertain one k extreme one k intermediate	$\bar{r}^{(1)}, \bar{r}^{(2)}, \bar{s}^{(1)}\ \bar{s}^{(2)}$, uncertain Both k's extreme	$\bar{r}^{(1)}, \bar{r}^{(2)}, \bar{s}^{(1)}, \bar{s}^{(2)}$ uncertain one k extreme one k intermediate	

one matrix). In seven of the cases, one is afforded at least two unambiguous vectors (one each from $\bar{s}$ and $\bar{t}$), and in the eight other cases at least two of the vectors derived from the extreme values of the matrix elements of $\tilde{K}$ correspond to the correct spectra. In any event, the range of possible spectra and elution vectors consistent with the data can be readily displayed (15,16).

Factor analysis has been applied to GC-MS data by a number of workers (17-25). The most obvious application of the technique is determining the number of components present in a "fused" peak. Pioneering work by Rogers' group (17,18) showed the possibility that two compounds with sufficiently different mass spectra could be readily detected, even when their retention times differed by less than one half standard deviation (for gaussian peaks) and the signal to noise ratio was only 10/1. With a better signal to noise ratio, a minor component as small as ≈5% could be reliably found (23). A very interesting study by Isenhour and co-workers (19) shows the benefits of factor analysis. One of their studies involved recording the intensities of 18 m/e values for seven cyclohexane/hexane mixtures. When this data was analyzed, three factors emerged, which indicated that three components were present. Careful examination of the data revealed that the third component was due to the molecular nitrogen background arising from air leaking into the instrument. Elimination of the 28 m/e channel gave only two factors as originally expected.

Surprisingly, it was not until 1981 that attempts were made to extract the individual mass spectra from the data. Sharaf and Kowalski (24) used principal component analysis (Equations 16) to determine the number of components contributing to a given peak. They then transformed the eigenvectors of the mass spectral space to be all-positive. They derived a graphical interpretation to provide the range of solutions consistent with the data. Spectra were then presented together with the ambiguity of the intensity of each of the mass channels. As is predicted from theory (15) some of the intensities of mass channels were found to be very ambiguous while others were found to have virtually no uncertainty. Clearly only these latter should be used in a spectral searching program.

In a second paper on the factor analysis of GC-MS data (25), Sharaf and Kowalski concentrated on determining the elution vectors of the individual components contributing to a fused peak. In favorable cases, where there were unique masses in the mass spectra of the overlapping components, the peak profiles could be determined with excellent certainty. Data was presented to show the possibility of detecting and identifying two components even if their retention times were identical - all that is needed is some difference in their peak profiles.

An alternative approach to principal component analysis has been described by Knorr, Thorsheim and Harris (26). These authors start with a linear model for GC-MS data as expressed by

Equation 13. However, instead of decomposing the data matrix into sets of orthogonal vectors (principal components), they assume a specific functional form for the retention vectors, $\bar{r}^k$:

$$r_i^k = \frac{1}{2\pi\sigma}\exp\left\{-\frac{1}{2}\left[\left(t_i-t_0^k\right)/\sigma^k\right]^2\right\}\exp\left(-t_i/\tau\right) \qquad [18]$$

and

$$\sigma^k = t_0^k\, N^{-\frac{1}{2}} \qquad [19]$$

with N being the number of theoretical plates. In this analysis, the parameters to be determined are the number of components, and the values of $\sigma, t_0, \tau, \alpha^k$, and the $\bar{s}^k$ for each. Since there are less parameters than data points, the method of least squares or maximum likelihood can be used to obtain estimates for the parameters. This technique appears to work extremely well for real data consisting of mixtures of three components when overlaps are exceedingly severe. It is important to realize that this scheme assumes ideal behavior in the chromatogram, i.e., that all the peak shapes are known to obey the same rules, i.e., where one has an ideal chromatogram. However, if this assumption does hold, the ambiguity in the results obtained by the Harris technique is greatly reduced compared to the factor analysis technique. Moreover, the former algorithm works quite well on systems of greater than two components.

Chen and Hwang (21) studied the application of factor analysis to GC-MS peaks with three components contributing. While it is certainly straight forward to obtain the three sets of eigenvectors, the difficulty is to find a transformation to spectral-space. Chen and Hwang give a graphical method for determining the possible range of mass spectral vectors. For a three component system the constructed spectrum can be represented in a three dimensional eigen (cartesian) space as follows:

$$\bar{s} = \bar{u}^1\cos\theta + \bar{u}^2\sin\theta\cos\phi + \bar{u}^3\sin\theta\sin\phi \qquad [20]$$

where θ and ϕ are the polar angles for the spherical coordinates of $\bar{s}$, and the $\bar{u}$ vectors are the eigenvectors. Then a plot is made in θ-ϕ space of all of the non-negative spectra transformed from A. The desired component spectra will lie on the vertices of the "feasible" region. In favorable cases, with minimal component overlaps, there will be three vertices and the three spectra will be found here. In cases where more overlap is present there will be "spurious vertices", and thus the solution will be ambiguous. Using this procedure, the investigators successfully reconstructed the mass spectra of 2 and 3 component mixtures. As expected, the result for the 3 component mixture was less satisfactory due to the greater degree of overlap.

Factor Analysis has also been applied to thin layer and liquid chromatography data. Gianelli et al (27) developed an instrument which could rapidly obtain a series of fluorescence spectra along the elution direction of a one dimensional thin layer chromatogram. These workers were able to resolve two highly overlapping components into their individual retention and spectral vectors. There are also obvious applications to HPLC now that multichannel absorption detectors are commercially available (28), and this aspect of the problem is being actively pursued by Kowalski (29).

We now come to the most difficult (and prevalent!) analytical situation where one has a mixture of unresolved components in the chromatogram, some of which are known and some of which are unknown. In this case, one wishes to (a) quantitate the knowns and (b) obtain the spectra and retention profiles of the unknowns for qualitative identification. We believe that "Rank Annihilation", developed by Davidson (30), is an excellent approach to performing this type of analysis. The technique can be described qualitatively as follows. The data is assumed to be represented by Equation 13. Thus, the rank of $\underset{\sim}{M}$ is equal to the number of components, both knowns and unknowns. Now, suppose that we have for the kth component, a calibration matrix $\underset{\sim}{N}^k$. Then, if we subtract the correct amount of $\underset{\sim}{N}^k$ from $\underset{\sim}{M}$, the remainder matrix, $\underset{\sim}{M}^{rem}$, will have a rank of exactly one less than the original matrix, i.e.:

$$\underset{\sim}{M}^{rem} = \underset{\sim}{M} - \beta \underset{\sim}{N}^k \quad [21]$$

In practice, because we do not know the amount of $\underset{\sim}{N}^k$ to subtract and because of the presence of noise in the data, we will have to try many values of β until the lowest meaningful eigenvalue of $\underset{\sim}{M}$ is at a minimum. Direct computation by this scheme is expensive because many diagonalizations of $\underset{\sim}{M}$ must be carried out to find the minimum of the lowest eigenvalue. We can avoid this problem by working in a reduced basis set for $\underset{\sim}{M}$ consisting of the first q eigenvectors of $\underset{\sim}{M}\underset{\sim}{M}^T$ or $\underset{\sim}{M}^T\underset{\sim}{M}$ such that $q \geqslant p$, the number of components. Thus, for a 3 component mixture, we can usually work in a reduced space so that only a 5x5 matrix or less is repeatedly diagonalized. Originally, rank annihilation was applied to multicomponent fluorescence data (30,31). The data consisted of measurements of the fluorescence intensity as a function of both excitation and emission wavelengths and has the same form as Equation 13 (11). Such data represents a "worst case" situation because for each component there are multiple peaks in both dimensions of the matrix. Thus the chances for overlap and ambiguity are much greater than for chromatography data.

Rank Annihilation has been applied to the problem of multiple overlapping species in Liquid Chromatography by McCue and Malinowski (34). These authors attempted to make a quantitative determination of ethylbenzene in the presence of *o*-

and p-xylene, using two dimensional data sets consisting of absorbance measurements as a function of retention time and spectral wavelength. The spectra and retention times of the components overlap to a high degree, as shown in Figure 5A and 5B. As tests, two mixtures of the three components were run; elution profiles are shown in Figures 6A and 6B, and the amounts of each are given in Table II. The analysis of ethylbenzene in the mixture shown in Figure 6A represents a particularly difficult case, because there is no spectral wavelength or retention time where ethylbenzene is detectable or elutes by itself. Nevertheless, when the two dimensional data matrices consisting of samples taken at 7 time intervals each examined at 24 wavelengths between 257 and 280 nm, were eigen-analyzed, their rank was clearly three, i.e. the first three eigenvalues of the data matrix times its transpose were much larger than the succeeding eigenvalues. Figure 7 shows the process of rank annihilation for estimating the quantity of ethyl benzene in the run of Figure 6A. As expected, when the correct amount of ethyl benzene is determined, the size of the third eigenvalue is at a minimum, indicating that the rank of the matrix has decreased to two. In Table II are presented the results of a rank annihilation analysis of ethyl benzene, showing excellent quantitative agreement with the known concentration. Obviously, this technique can be readily applied to multidetector chromatography.

Table II. Comparison of Known Weight Fractions of Mixtures with Those Determined by Rank Annihilation (RA) for Ethyl Benzene.

Solute	Wt.Fr.in Soln.1 prepared	RA	Wt.Fr.in Soln.2 prepared	RA
Et-benzene	.249±.004	.239	.599±.007	.594
o-xylene	.505±.006	----	.203±.003	----
p-xylene	.247±.004	----	.198±.003	----

Chromatography with Multidimensional Detectors

In this section, we will consider the use of multidimensional spectrometers as chromatography detectors. In multidimensional spectroscopy, the spectral intensity is a function of more than one spectral parameter. For example, detection by means of fluorescence is inherently two dimensional because the observed intensity is a function of two variables, wavelength of emission and wavelength of excitation (15). Other examples of two dimensional spectroscopy are: MS-MS (35) and the various types of two dimensional NMR experiments (36). Appellof and Davidson

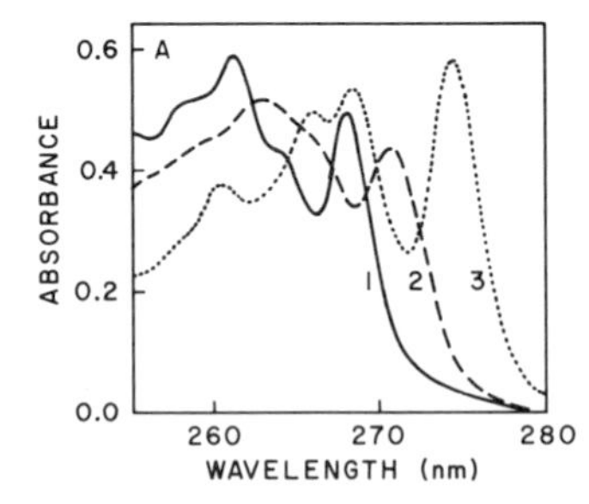

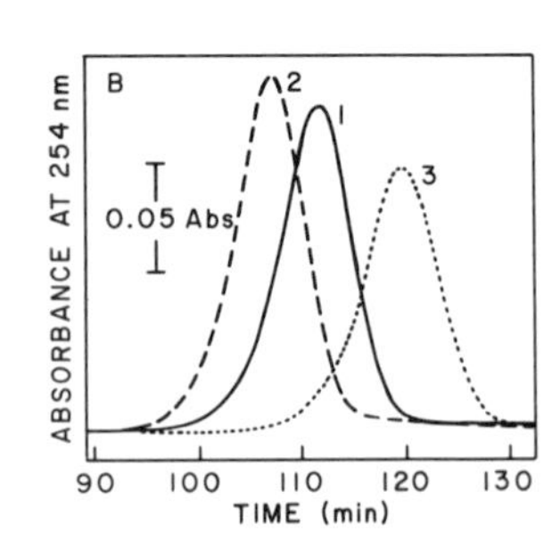

Figure 5. A. Absorbance spectra of (1) ethyl-benzene, (2) o-xylene and (3) p-xylene. **B**. Retention profiles for (1) ethyl-benzene, (2) o-xylene and (3) p-xylene.

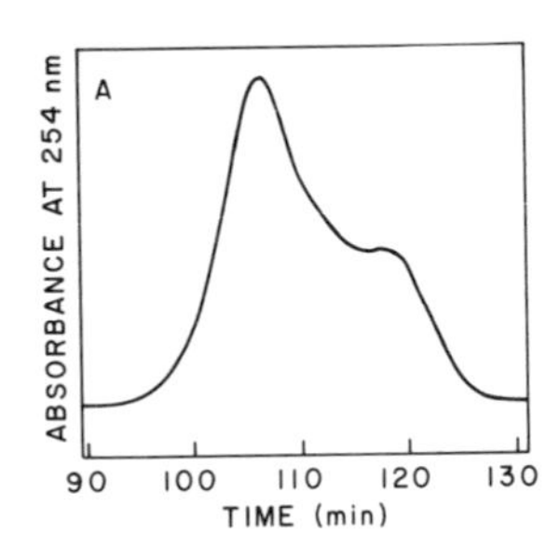

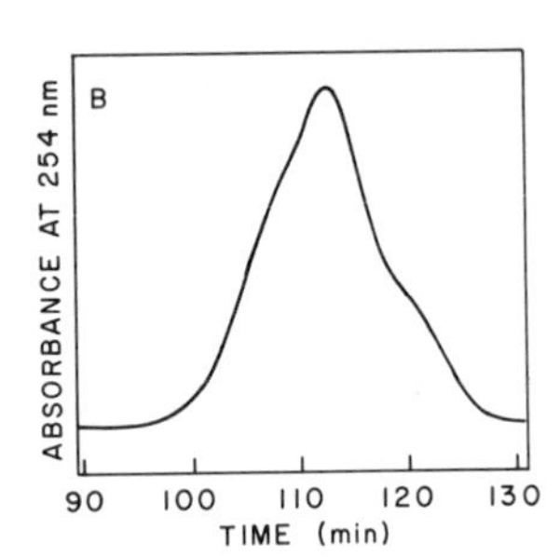

Figure 6. A. Retention profile for the first mixture. **B**. Retention profile for the second mixture.

have considered various strategies for analyzing the data which results from video fluorometric monitoring of liquid chromatography effluents (37,38). The three dimensional data array is assumed to have the following form:

$$\underset{\sim}{T} \equiv \left[t_{ijk}\right] \; ; \; t_{ijk} = \sum_{\ell=1}^{P} \alpha^{\ell} x_i^{\ell} y_j^{\ell} r_k^{\ell} \qquad [22]$$

or

$$\underset{\sim}{T} = \sum_{\ell=1}^{P} \alpha^{\ell} \bar{x}^{\ell} * \bar{y}^{\ell} * \bar{r}^{\ell} \qquad [23]$$

where the third-order tensor $\underset{\sim}{T}$ is the set of intensity measurements as a function of wavelength of emission indexed by i, a function of wavelength of excitation indexed by j, and a function of time, indexed by k. The sequenced set $\bar{x}^{\ell} \equiv \{x_i^{\ell}\}$ represents the normalized emission spectrum of the ℓth component; $\bar{y}^{\ell} \equiv \{y_j^{\ell}\}$, the normalized excitation spectrum, and $\bar{r}^{\ell} \equiv \{r_k^{\ell}\}$, the normalized retention vector. We have assumed that the three variables are independent and linearly additive so that the entire data set is expressed as the sum of the individual components which are, in turn, expressed as the Kronecker products of their emission, excitation and retention vectors.

In the case where all of the components are knowns, for which standard chromatograms are available, it is quite straightforward to use the method of least squares to obtain estimates for the amounts of each component. The generalization of Equations 4 and 15 to the case of two dimensional chromatography is trivial and will not be repeated here. In the case where none of the components are knowns, it is nevertheless possible to estimate the number of components, as well as the spectra and retention times of each (37). Here, a linear decomposition of the array cannot be accomplished by means of eigenanalysis, as before, and a non-linear least squares method must be used. However, if a decomposition can be found, it can be shown to be totally unique, unlike the case of a two-dimensional matrix (38). Essentially, we attempt to minimize the quantity R, defined by

$$R^2 = \sum_{ijk} \left[t_{ijk} - \sum_{\ell=1}^{P} x_i^{\ell} y_j^{\ell} r_k^{\ell}\right]^2 \qquad [24]$$

where the quantity α, which has no meaning in the absence of calibration, has been absorbed into the definition of $\bar{x}$, $\bar{y}$, and $\bar{r}$. Unfortunately, the parameters $\bar{x}$, $\bar{y}$, $\bar{r}$ and P must be obtained by iterative non-linear methods, which are not only computationally expensive, but require a good starting estimate. However, as Appellof and Davidson have shown, it is possible to obtain a reasonable basis set for each of the three dimensions

$(\bar{x},\bar{y},\bar{r})$ of the array from the eigenvectors of the covariance matrix of each dimension. We keep only that number of eigenvectors which are associated with sufficiently large eigenvalues. Using these basis sets greatly reduces the computational difficulties of the problem, as well as provides an excellent approximation to the number of components in the mixture.

In Figure 8 we show the results of the decomposition of a two component mixture, 9-methylanthracene and benzo(e)pyrene, using video fluorometer detection of the HPLC effluents. Comparison of the estimated excitation and emission spectra with those of the pure compounds shows excellent agreement. As can be seen, the spectral overlaps are very great, as are the elution profile overlaps (less than one standard deviation, not shown). In Figure 9 are the results of a decomposition of a three component mixture containing the two components of Figure 8 plus benzo(a)pyrene. The overlaps are even more severe, yet the calculated spectra show acceptable agreement with the true spectra.

The techniques of factor analysis may also be performed in three dimensions (33). Again, the basic idea is the same as for one dimensional chromatography, expressed by Equation 22-23. As above, we work in a basic set consisting of the eigenvectors of the covariance matrix for each of the three dimensions of $\underset{\sim}{T}$. This renders the analysis computationally feasible. A rigorous sensitivity analysis of the technique shows that the three dimensional analysis will always have less overlap (less ambiguity) and is therefore more likely to provide better accuracy than a two dimensional analysis. This assertion has been verified for simulated data wherein highly overlapping spectra and elution profiles were used. We will look forward to seeing the technique applied to real data.

Multidimensional Chromatography with Multichannel Detection

In multidimensional chromatography, the separation of a particular component is a function of its retention time along two or more dimensions. Two practical implementations of two dimensional chromatography are two dimensional TLC (39) and two dimensional gel electrophoresis (40). When these separation dimensions are combined with one or more spectroscopic dimensions a very powerful analytical system results (41). We have recently constructed a single beam imaging spectrophotometer which is capable of recording the reflectance or transmittance spectrum of up to 64,000 positions on a two dimensional thin layer plate within a few minutes. A block diagram of the instrument is given in Figure 10. The TLC plate is uniformly illuminated with monochromatic light of a selected wavelength. Using a tungsten lamp and monochromator, the transmission or reflectance image of the plate is recorded by a digital imaging device, previously described (42). The wavelength is changed and a new image

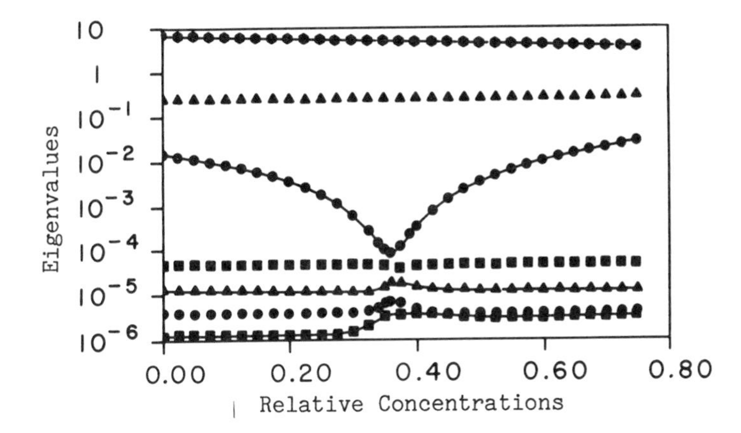

Figure 7. Eigenvalues as a function of the relative concentration of the ethyl-benzene standard subtracted from diluted solution 1.

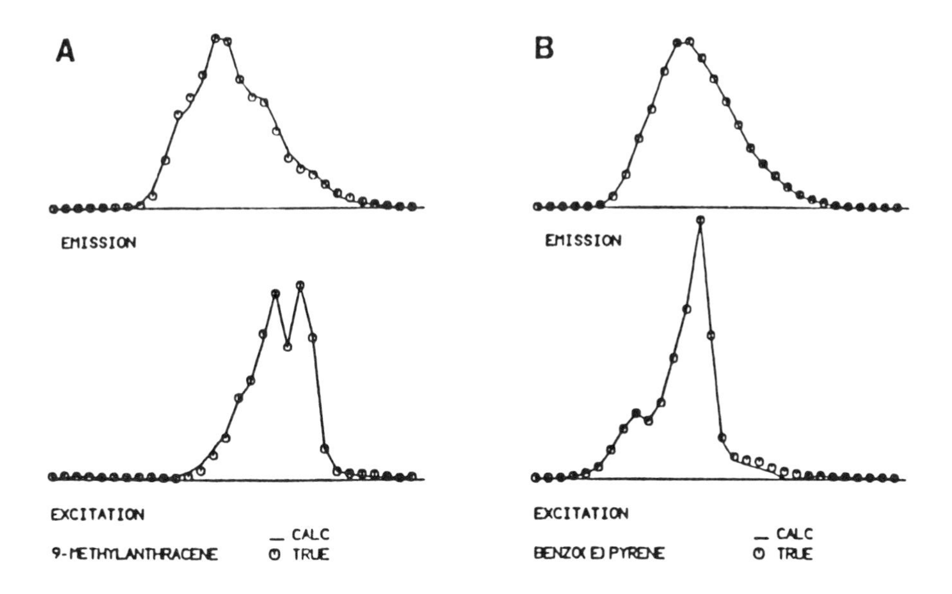

Figure 8. Comparison of true emission and excitation spectra (⊙) to those calculated (solid line) for a sample containing 9-methyl anthracene (A) and benzo(e)pyrene (B).

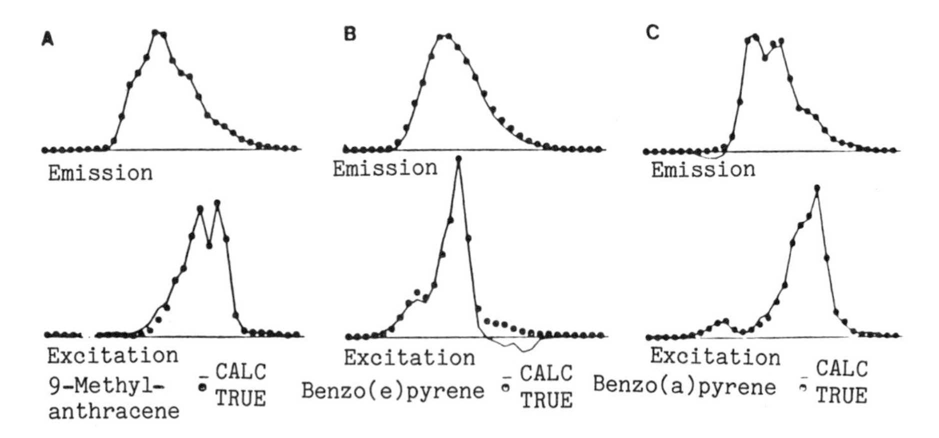

Figure 9. Comparison of true emission and excitation spectra (⊙) to those calculated (solid line) for a sample containing 9-methyl anthracene (A), benzo(e)pyrene (B), and benzo(a)pyrene (C).

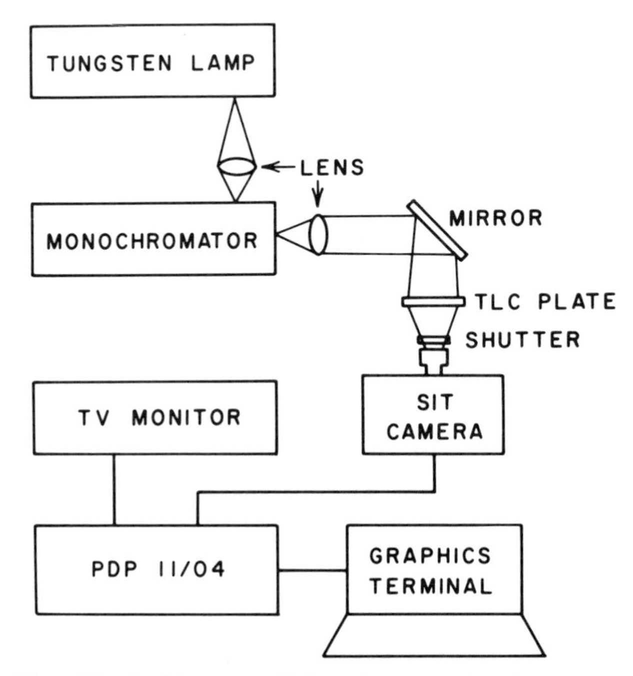

Figure 10. Block diagram of imaging spectrophotometer.

recorded and so on. The result is a "stack" of images stored on disk and available for recall and processing by various techniques. It can be shown that the data from this instrument can be represented by a three dimensional array which consists of the outer product of two elution vectors and one spectroscopic vector; i.e., it has the same form shown in Equations 22-23 for video fluorometric detection of HPLC effluents.

The multichannel imaging spectrophotometer has been applied to the analysis of a mixture of three porphyrines which are only partially resolved on a two dimensional plate. Figure 11 shows several absorbance images of the plate taken at various wavelengths for the mixture. Especially striking is the degree of differential enhancement of the selected (compound) spots as a function of wavelength of illumination. In 11A, the plate is imaged at 420 nm where all of the porphyrins absorb. The next three images were taken at 523 nm, 554 nm and 594 nm; H_2TPP absorbs strongly at 523 nm, PdEP at 554 nm, and ZnTPP at 594 nm. In 11B, then, the main contour peak is due to H_2TPP, while there is a weaker peak to the left of it due to weak absorption by PdEP. Likewise, in 11C the left PdEP peak is strong and the right H_2TPP peak somewhat diminished. In 11D, only the ZnTPP is seen. Quantitative analysis of this mixture is summarized in Table III using rank annihilation and least squares.

Figure 12 illustrates the result of qualitative analysis of the two dimensional chromatogram by decomposing it according to the algorithm detailed in (37). The input data, assumed to be of the form of Equation 22, are shown as a "stack" of images in Figure 12A, typical examples of which were displayed in Figure 11. The results from the decomposition of these data into their three components is illustrated in Figure 12 B-D. The absorption spectrum and the retention profiles along the x and y axes are all shown for each of the three components. The spectra obtained superimposed nearly exactly on those of the pure components.

Table III. Two Dimensional Chromatography Data Analyzed by Rank Annihilation, RA, and Least Squares, LS.

Species	Present	RA	LS
H_2TPP	1.40μg	1.57μg	1.69μg
ZnTPP	0.91μg	0.86μg	0.73μg
PdEtio	1.37μg	1.44μg	1.46μg

Conclusions

Numerical deconvolution appears to be a viable approach to obtaining higher resolution in chromatography. In the case where

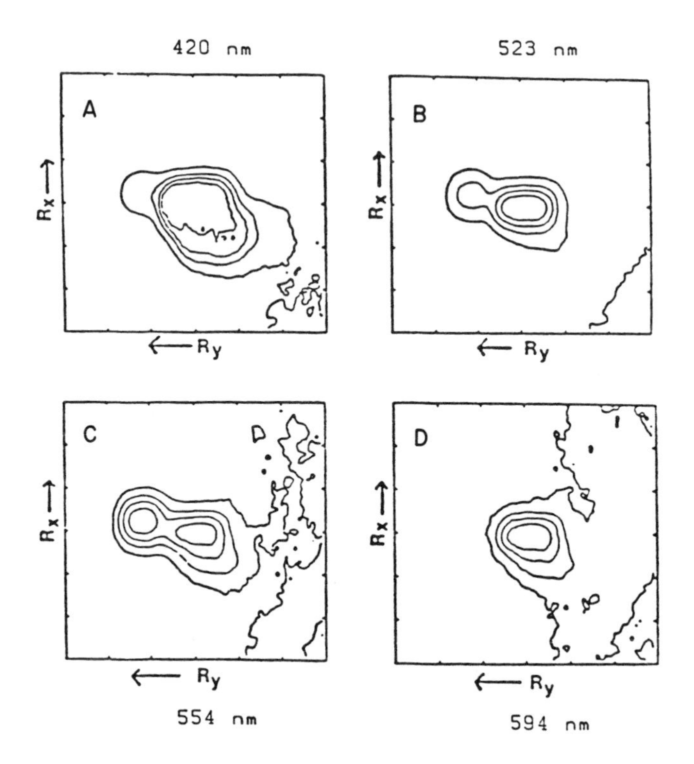

Figure 11. Two-dimensional chromatography contour plots of a mixture of H_2TPP, ZnTPP, and PdEP at selected absorption wavelengths. **A**. 420nm, **B**. 523nm, **C**. 554nm, and **D**. 594nm.

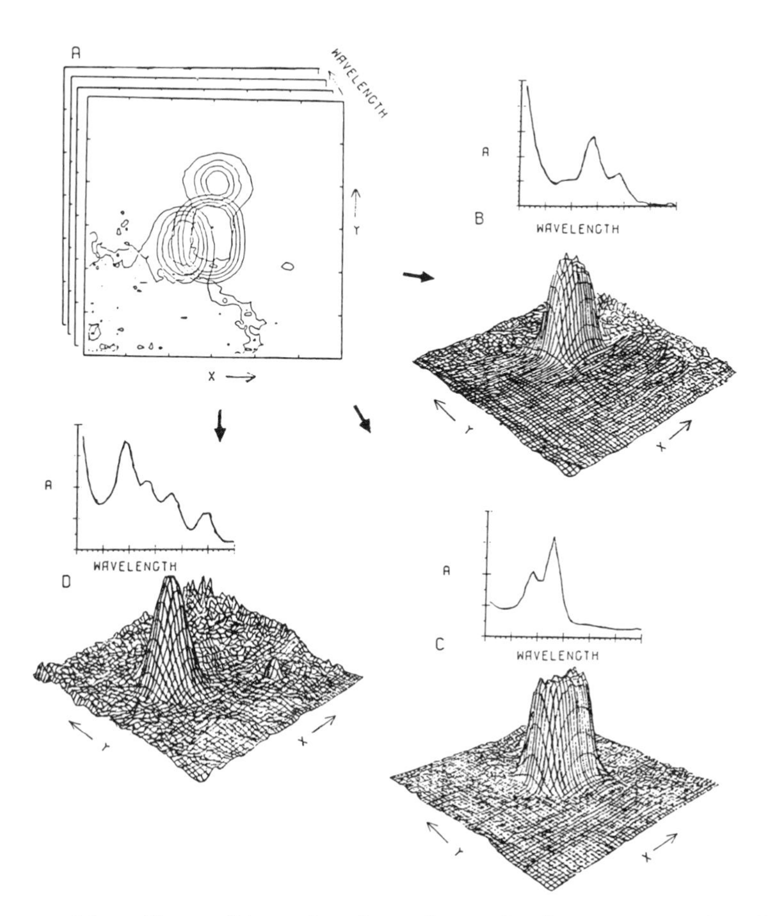

Figure 12. Three-dimensional rank analysis. **A**. Stack of all two-dimensional chromatography images as a function of wavelength. **B–D**. Individual spatial-spectral components (B = PdEP; C = H_2TPP; D = ZnTPP); each three-dimensional plot is a spatial representation of the associated compound (spectrum) obtained by factorization of the stack of chromatography images.

the chromatographic effluents are detected with a single channel detector, we are simply trying to cancel the effect of the broadening mechanisms. The method we have proposed shows a great deal of promise for achieving super-resolution and estimating the number of components under a fused peak. However, it does have two drawbacks common to all techniques which seek to solve a convolution equation of this kind: (a) it requires an extremely good signal/noise ratio, and (b) it requires some assumption about the peak shape. Fortunately, in a sufficient number of cases, there will be an excellent signal/noise ratio and the peak shapes will be regular and predictable. Under the most favorable circumstances it seems quite feasible to deconvolute peaks which are within one half of a standard peak width of each other. This is equivalent to a four times increase in resolution or a sixteen times increase in column length. Unfortunately, the number of fully resolved peaks will go up only as the square root of the resolution (1) so that one dimensional techniques will not improve our analytical capabilities as much as we would like.

In contrast, chromatography schemes which involve multiple detectors or a multichannel spectrometric detector will take us much further toward our goal. The data set from such an instrument is now two dimensional, and the overall resolution will be the product of the individual resolutions; thus the likelihood of overlap is smaller to begin with. Moreover, in favorable cases where the contributions of the individual components is linearly additive, powerful theorems of linear algebra may be used to analyze the data. In contrast to one dimensional data, we now have a rigorous method (15,18,19) for estimating the number of components in the mixture (Rank Determination), a method (30-32,34) for quantitating knowns in an unknown, interfering background (Rank Annihilation), and a technique (17,24,25) for estimating the spectra and retention times of each of the unknowns (Factor Analysis). In favorable cases, at least as shown by computer simulations, it seems quite feasible to resolve components which are only one-tenth of a standard deviation apart.

Finally, we considered the use of three dimensional analytical techniques such as the GC-MS-MS and video fluorometric monitoring of HPLC effluents. As with two dimensional techniques, we can use Rank Analysis, Rank Annihilation and Factor Analysis. However, in this case the data can be shown to be unambiguously decomposible, even when there is severe overlap among components.

Thus, there seems to be a great deal of promise for numerical deconvolution in improving chromatographic resolution. All that now remains is to put these techniques into the hands of practicing analytical chemists where they can be used on a routine basis.

Acknowledgment

Useful discussion with Ernest Davidson and Thomas Bellus is gratefully acknowledged. This research was supported in part by NIH grants GM-22311 and GM-26935, and by DOE contract number Y33-S-44-22462.

Literature Cited

1. Hirschfeld, T., Anal. Chem. **1980,** 52, 297A-312A.
2. Rosenthal, D., Anal. Chem. **1982,** 54, 63-66.
3. Davis, J.M.; Giddings, J.C., Anal. Chem., **1983,** 55, 418-424.
4. Bevington, P.R., "Data Reduction and Error Analysis in the Physical Sciences," McGraw-Hill, New York, **1969.**
5. Rosenbaum, M.; Hanid, V.; Komers, R., J. Chromatogr., **1980,** 191, 157-167.
6. Gold, H.S.; Rechsteiner, C.E.; Buck, R.P., Anal. Chem., **1976,** 48, 1540-1546.
7. Nomura, H.; Shinobu, K.; Miyahara, Y., Appl. Spectros. **1979,** 33, 248-253.
8. Maddams, W.F., Appl. Spectrosc. **1980,** 34, 245-267.
9. Gillete, P.C.; Lando, J.B.; Koenig, J.L., Appl. Spectrosc. **1982,** 36, 401-404.
10. Grinvald, A.; Steinberg, I.Z., Anal. Biochem. **1974,** 59, 583-598.
11. Warner, I.M.; Davidson, E.R.; Christian, G.D., Anal. Chem. **1977,** 49, 2155-2159.
12. Frieden, B.R., in "Picture Processing and Digital Filtering"; Huang, T.S., Ed.; TOPICS IN APPLIED PHYSICS, Vol. 6, Springer-Verlag: New York, **1975;** Chapter 5.
13. Lawson, C.L.; Hanson, R.J., "Solving Least Squares Problems"; Prentice-Hall: Englewood Cliffs, NJ, **1974, 1981;** Chapter 23.
14. Malinowski, E.R.; Howery, D.G., "Factor Analysis in Chemistry"; Wiley-Interscience: New York, **1980.**
15. Warner, I.M.; Christian, G.D.; Davidson, E.R.; Callis, J.B., Anal. Chem. **1977,** 49, 564-573.
16. Aartsma, T.J.; Gouterman, M.; Jochum, C.; Kwiram, A.L.; Pepich, B.V.; Williams, L.D., J. Am. Chem. Soc. **1982,** 104, 6278-6283.
17. MacNaughton, D.; Rogers, L.B.; Wernimont, G., Anal. Chem. **1972,** 44, 1421-1427.
18. Davis, J.E.; Shepard, A.; Stanford, N.; Rogers, L.B., Anal. Chem. **1974,** 46, 821-825.
19. Riter, G.L.; Lowry, S.R.; Isenhour, T.L.; Wilkins, C.L., Anal. Chem. **1976,** 48, 591-595.
20. Malinowski, E.R. and McCue, M., Anal. Chem. **1977,** 49, 284-287.
21. Halket, J. McK., J. Chromatogr. **1979,** 175, 229-241.
22. Chen, J.-H.; Hwang, L.-P., Anal. Chim. Acta **1981,** 133, 271-281.

23. Woodruff, H.B.; Tway, P.C.; Cline Love, L.J., Anal. Chem. **1981**, 53, 81-84.
24. Sharaf, M.A.; Kowalski, B.R., Anal. Chem. **1981**, 53, 518-522.
25. Sharaf, M.A.; Kowalski, B.R., Anal. Chem. **1982**, 54, 1291-1296.
26. Knorr, F.J.; Thorsheim, H.R.; Harris, J.M., Anal. Chem. **1981**, 53, 821-825.
27. Gianelli, M.L.; Callis, J.B.; Andersen, N.H.; Christian, G.D., Anal. Chem. **1981**, 53, 1357-1361.
28. Miller, J.C.; George, S.A.; Willis, B.G., Science, **1982**, 218, 241-246.
29. Kowalski, B.R., private communication.
30. Ho, C.-N.; Christian, G.D.; Davidson, E.R., Anal. Chem. **1978**, 50, 1108-1113.
31. Ho, C.-N.; Christian, G.D.; Davidson, E.R., Anal. Chem. **1980**, 52, 1071-1079.
32. Ho, C.-N.; Christian, G.D.; Davidson, E.R., Anal. Chem. **1981**, 53, 92-98.
33. Appellof, C.J.; Davidson, E.R., Anal. Chim. Acta **1983**, 146, 9-14.
34. McCue, M.; Malinowski, E.R., J. Chromatogr. Sci. **1983**, 21, 229-234.
35. Yost, R.A. and Enke, C.G., Anal. Chem. **1981**, 51, 1251A-1264A.
36. Bax, A. "Two Dimensional Nuclear Magnetic Resonance Spectroscopy in Liquids," Delft University Press, Delft Holland, **1982**.
37. Appelloff, C.J.; Davidson, E.R., Anal. Chem. **1981**, 53, 2053-2056.
38. Appelloff, C.J., Ph.D. thesis, University of Washington, Seattle, Washington, **1981**.
39. Touchstone, J.C.; Sherma, J. "Densitometry in Thin Layer Chromatography," John Wiley and Sons: New York, **1979**.
40. Lester, E.P.; Lemkin, P.F.; Lipkin, L.E., Anal. Chem. **1981**, 51, 391A-404A.
41. Gianelli, M.L.; Burns, D.H.; Callis, J.B.; Christian, G.D.; Andersen, N.H., Anal. Chem. **1983**, 55, in press.
42. Johnson, D.W.; Gladden, J.A.; Callis, J.B.; Christian, G.D., Rev. Sci. Inst. **1979**, 50, 118-126.

RECEIVED January 10, 1983

13

The Role of Ultrahigh Resolution Chromatography in the Chemical Industry

H. MICHAEL WIDMER and KARL GROLIMUND

Department of Analytical Research, CIBA-GEIGY Ltd., CH-4002 Basel, Switzerland

Trace investigations are routine activities of the industrial analyst, who attempts to solve problems in quality control, product research and development, and supports production, safety and marketing managers. Challenged by corporate specifications and requirements from authorities, his goal is to find economic methods of low detection limits and improved separation efficiencies. Instrumental techniques are increasingly combined with chromatographic methods in which low effluent flows and high separation power are pre-requisites. The number of impurities present in a sample at ppb level is orders of magnitudes larger than that present at ppm levels. Accordingly, there is a need for ultrahigh resolution chromatography techniques. Theoretical and practical aspects of ultrahigh resolution gas chromatography and high performance liquid chromatography are discussed. Examples from capillary GC and microbore HPLC are presented.

Analytical chemistry is an extremely dynamic branch of science and the industrial analyst is constantly faced with problems associated with methodological changes. Only about 20 years ago trace investigations in the ppm range became accessible, today the ppb range is under attack. This evolution is not simply a matter of extending analytical experience to a field of higher sensitivity and lower detection limits, but it involves the introduction and development of new separation techniques and new detection methods. Furthermore it is accompanied by an appropriate change of mind and mental attitude.

Ultrahigh Resolution Chromatography in Industry

In our days the industrial analysts fight on two frontiers at the

0097-6156/84/0250-0199$06.75/0

same time. On one side they are interested in the development of low detection limit systems and on the other they strive to create the basis for high resolution separation techniques.

Both problems are interrelated and one cannot solve one problem without touching the other. The lower the concentration of a component in a sample the larger is the number of interfering substances in this sample. Matrix effects are the most persistent problems in the daily work of the industrial analyst concerned with trace and ultratrace investigations. This is demonstrated in "Figure 1". It represents a set of chromatograms in which gasoline is analyzed by capillary GC. The difference of the four chromatograms lays in the change of sensitivity of the flame ionization detector setting. The top chromatogram is taken with an attenuation of 4096, the last with 1. The figure shows that there is one component present at a concentration higher than 10 %, however, there are 10, 43 and 105 sample components at concentrations higher than 1 %, 1000 ppm and 100 ppm, respectively("Figure 2"). This trend is a most general one and not specific for gasoline. Similar relations exist with cigaret smoke, polluted air, water and ocean pollution as well as industrial products.

There are several reasons for the industrial analyst to show an interest in ultratrace investigations and therefore he must cope with ultrahigh resolution chromatography.

In chemical industry there is a great awareness about the quality of the products and quality control has never been taken so seriously and consequently as in our days. Unwanted side effects of certain components, the presence of impurities with carcinogenic, mutagenic or teratogenic effects, corporate and governmental requirements force the quality control people to apply trace and ultratrace methods.

A significant drive for the need to lower detection limits stems from occupational safety and health considerations and requirements. Hazardous chemicals such as dimethyl sulfate may be used as educts in the synthesis of industrial goods, others such as epichlorohydrine or bis(chloromethyl)ether may be generated during a chemical process.

"Table I" summarizes the maximum permissable working concentrations of these substances in air. In Switzerland (CH) and Germany (GFR) they are called MAK-values, in the USA they are known as Threshold Limit Value-Time Weighted Average (TLV-TWA). CIBA-GEIGY established its own Permissible Internal Exposure Level (PIEL) for substances, such as bis(chloromethyl)ether, for which official levels are not available.

Monitors. There are several techniques to check and control hazardous material in the air of working areas. In recent years the permanent surveillance of ambient air has become an important issue of the safety and health people in industry and the analyst has to conform with these needs and provide the appropriate methods and instrumentation.

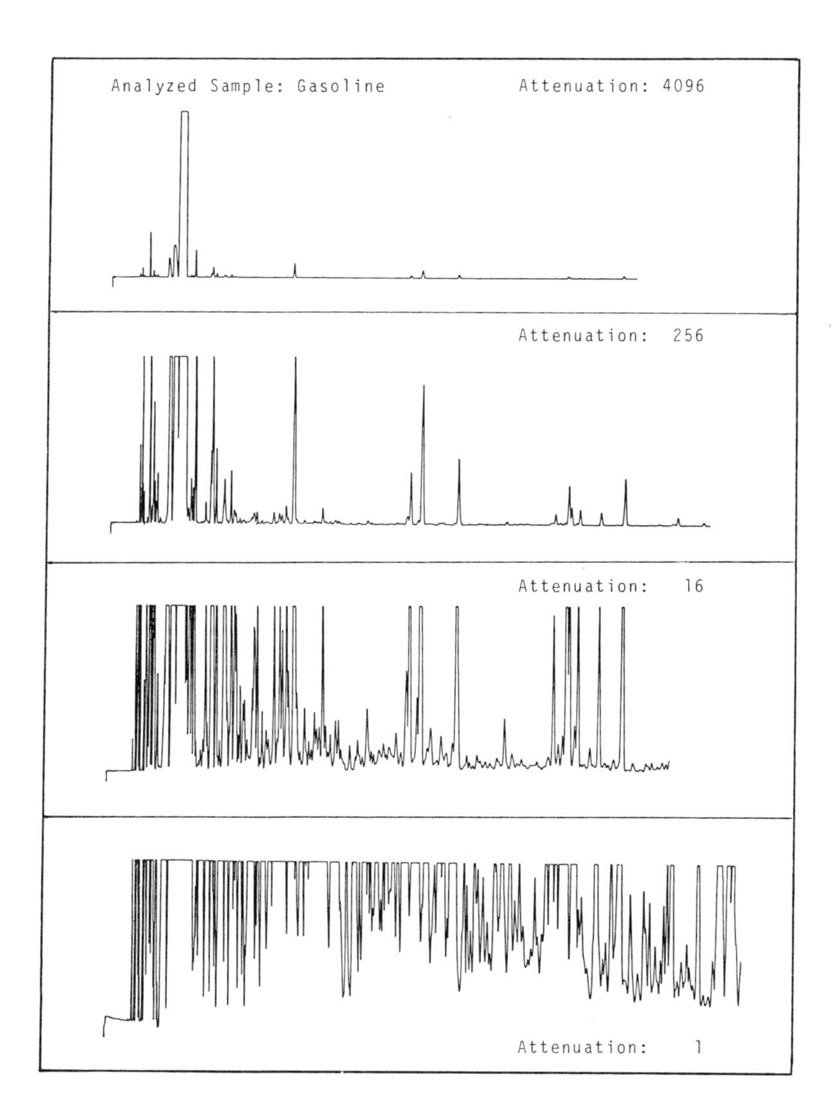

Figure 1. Gasoline analyzed by capillary GC (SE-54, 60 m).

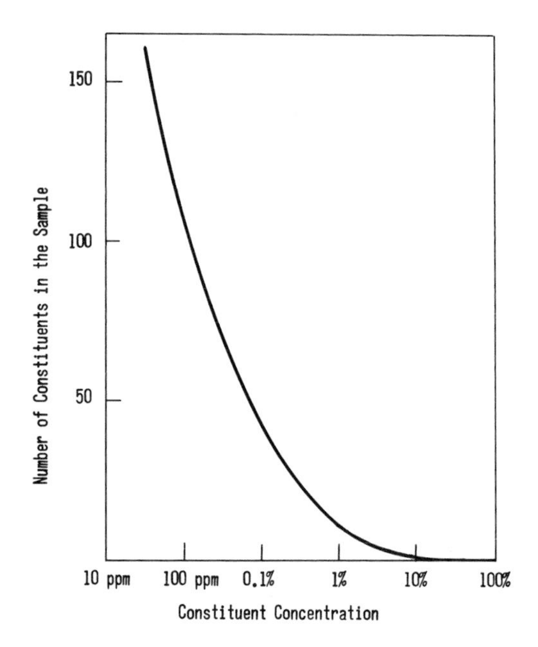

Figure 2. Plot of constituent number versus constituent concentration in gasoline.

Table I. Maximum Permissible Exposure Level for DMS, ECH and BCME

Species	Exposure Level				Monitor Detection Limit
	USA TLV-TWA	GFR MAK	CH MAK	PIEL	
DMS	100 ppb	10 ppb	10 ppb	-	2 ppb
ECH	2 ppm	3 ppm	5 ppm	-	100 ppb
BCME	1 ppb	-	1 ppb	500 ppt	200 ppt

In our company we developed a versatile and flexible monitor system shown in "Figure 3". This particular instrument is used for the surveillance of bis(chloromethyl)ether in ambient air.

Our aim was to design and construct a mobile automated system, performing survey analyses at a rate of 1-5 analyses per hour on the hour. "Figure 4" shows the operational principle of the instrument. It consists of compact module units, which may be arranged for specific applications and serve different needs (1, 2).

The air sample is trapped on an adsorbent, then thermally desorbed and gas chromatographed, followed by an appropriate detection system, such as FID, FPD, ECD or a mass spectrometer. Examples of detection limits are given in "Table I".

Theoretical Considerations

The chromatographic resolution is given by the column efficiency and the selectivity of the chosen separation system.

Efficiency. The efficiency may be described by "Equation 1"

$$E_B = \frac{t_{R_B}}{W_B} \qquad (1)$$

For packed columns the efficiency is given by the quality and particle size of the column packing. For wall coated capillaries the efficiency is a matter of the coating film properties. Furthermore, the efficiency depends on the injection mode, solvent effects, flow rate and column dimensions.

Generally, the column efficiency is expressed in terms of the theoretical plate number N or the height equivalent to a theoretical plate, H :

The plate number is

$$N = 16 \left(\frac{t_{R_B}}{W_B} \right)^2 = 16\ E^2 \qquad (2)$$

and the height equivalent to a theoretical plate.

Figure 3. Bis(chloromethyl)ether monitor.

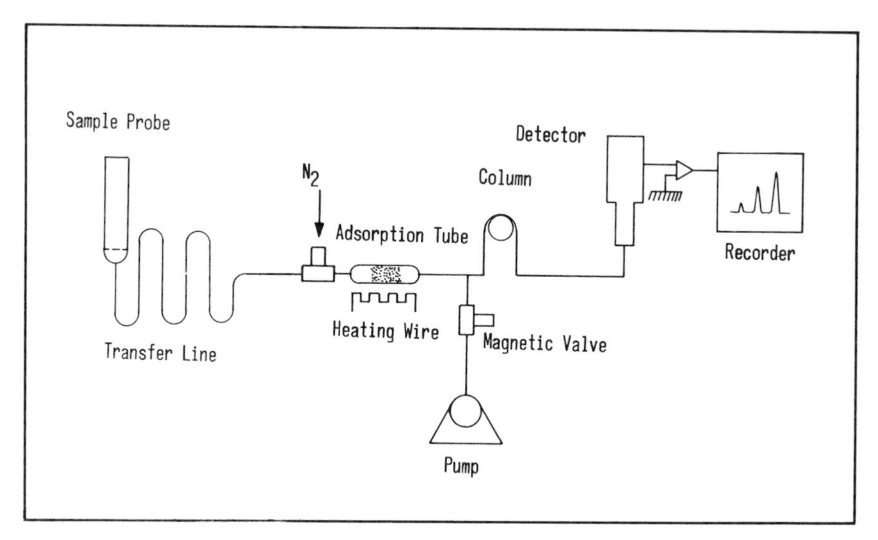

Figure 4. Schematic view of monitor (operational principle).

$$H = \frac{1}{N} = \frac{1}{16\ E^2} \qquad (3)$$

For a given column at optimum flow conditions the maximum efficiency and theoretical plate number can be calculated, as well as the minimum H.

"Table II" and "Table III" give some representative values for HPLC and capillary GC, respectively.

Table II. Efficiency of HPLC Columns with Different Packing Materials

Diameter of Packing Material in μm	10 cm Column E_{Max}	10 cm Column N_{Max}	20 cm Column E_{Max}	20 cm Column N_{Max}	H_{Min} mm
10	18	5,000	25	10,000	0.02
5	25	10,000	35	20,000	0.01
3	32	16,000	45	32,000	0.006

Golay (3, 4) introduced a simplified formula for the minimum height equivalent to a theoretical plate for wall coated capillaries, given in "Equation 4" (5)

$$H_{Min} = r \left[\frac{1 + 6k' + 11k'^2}{3(1 + k')^2} \right]^{1/2} \qquad (4)$$

Table III. Efficiency of Wall Coated Capillary Columns

Inner Diameter mm	H_{Min} mm	N_{Max}/m	N_{eff}/m	H_{eff} mm
0.10	0.084	11,880	8,250	0.121
0.20	0.168	5,950	4,130	0.242
0.30	0.252	3,960	2,750	0.364
0.40	0.337	2,970	2,060	0.485
0.50	0.421	2,380	1,650	0.606

The height equivalent to a theoretical plate may be calculated by a semi-empirical Van Deemter equation (6)

$$H = A + \frac{B}{\bar{u}} + C \cdot \bar{u} \qquad (5)$$

In HPLC the Van Deemter curves, relating H with the mean linear velocity of the mobile phase, are hyperbolic and the minimum H is obtained with relatively low linear velocity around 1 and 2 mm/s. A deviation from the optimum flow is connected with a pronounced increase of the height equivalent, according to the B and C terms of "Equation 5", respectively.

In SFC this increase is less dramatic, because the diffusion coefficient is 2 to 3 orders of magnitude larger than in HPLC. As a consequence the minimum H becomes almost independant of the flow rate over a large range of linear velocity. For 3 µm columns a linear flow velocity between 0.4 and 1.2 cm/s or 2.8 to 8.5 ml/min for a column of 4.6 mm inner diameter may be achieved with a plate number around 120,000 per meter.

Selectivity. The selectivity, characterizing the elution of two different substrates is described by "Equation 6".

$$S_{A,B} = \frac{t_{R_B} - t_{R_A}}{t_{R_B}} \qquad (6a)$$

It depends mainly on the difference of the interactions between the substrates and the carrier material and the mobile and stationary phases.

Introducing the separation factor and capacity factor one obtains

$$S_{A,B} = \frac{(\alpha - 1)}{\alpha} \left(\frac{k'_B}{1+k'_B} \right) \qquad (6b)$$

Resolution. The resolution is derived from the efficiency and selectivity according to "Equation 7"

$$R_s = E_B \cdot S_{A,B} \qquad (7a)$$

and therefore

$$R_s = \left(\frac{\sqrt{N}}{4}\right) \cdot \left(\frac{\alpha-1}{\alpha}\right) \left(\frac{k'_B}{1+k'_B}\right) \qquad (7b)$$

Conclusions. From the foregoing considerations we can make some conclusions about the possibilities to improve the chromatographic resolution. In industry efforts are undertaken to improve the efficiency as well as the selectivity. They are summarized in the following lists.

Ways to Improve the Selectivity :

- to enhance specific interactions of the substrate with the mobile phase by
  - gradient elution
  - ion pairing
  - precolumn derivatisation
  - modifiers in SFC
- to enhance specific interactions of the substrate with the stationary phase by
  - multidimensional chromatography
  - column switching
  - specific phases (such as in size exclusion chromatography or optical isomer separation)
  - precolumn derivatisation
- to choose specific detection systems (there is a particular need for new and improved HPLC detectors)
  - hyphenated systems
  - post column derivatisation

Ways to Improve the Efficiency :

- The efficiency depends on the column quality and therefore on
  - column packing
  - particle size
  - film properties of wall coated columns
  - column dimensions

Generally, the expert chromatographer relies on his own column packing techniques, because commercial packed columns for HPLC do not always provide optimum efficiency, but are less vulnerable to the treatment of less experienced chromatographers.

Practical Aspects

Chromatographic theories concerned with efficiency and selectivity are generally based on homologous compounds, for which similar temperature dependant properties are known. However, in chemical industry the analyst is rarely concerned with mixtures of homologous substances. In these samples the temperature effects may outweigh those of doubling the film thickness and column length of capillary columns.

The following capillary GC tests were performed with the Grob Test Mixture (7), listed in "Table IV". It represents a combination of constituents of variable polarity and reflects quite well the real world of industrial problems.

In "Figure 5" a comparison is made between isothermal and temperature programmed capillary GC with equal retention times for the last eluting component of the Grob Test Mixture. In order to achieve these conditions the isothermal chromatogram was run at 150°C and the temperature programmed chromatogram was started at 60°C for 1 min, followed by a temperature rate of 4°C/min, whereby peak A elutes at 197°C.

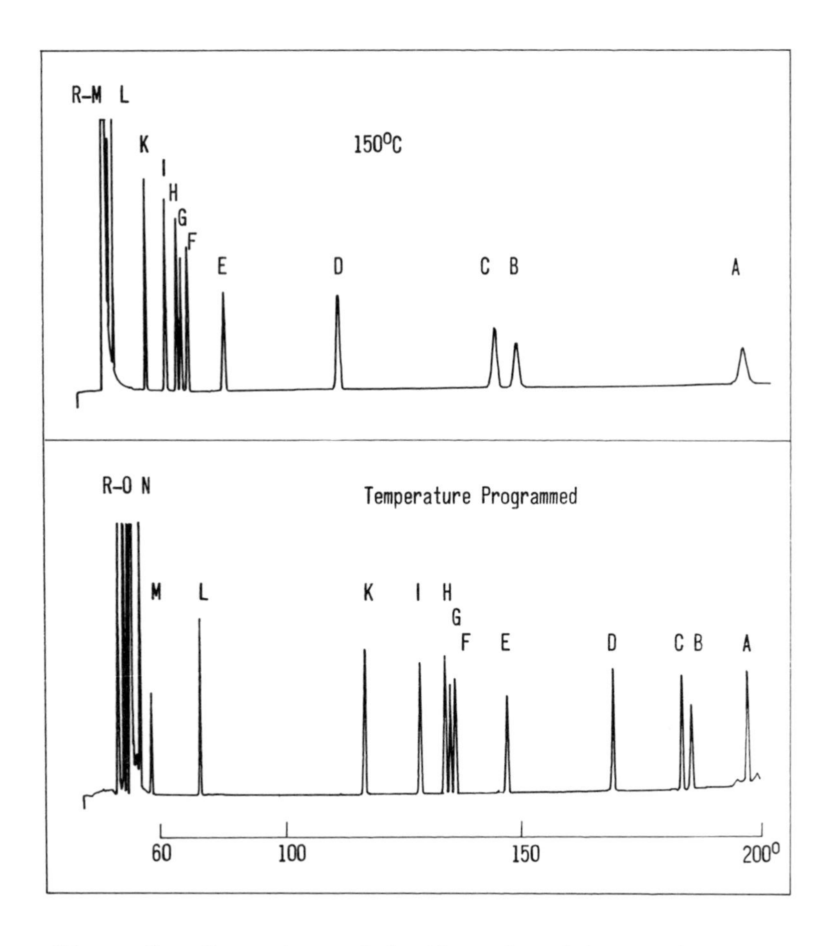

Figure 5. Comparison of isothermal and temperature programmed capillary GC of a test mixture.

Table IV. Grob Test Mixture

Compound	Chemical Formula	Abreviation	Peak Identification in "Figures 5, 6, 8 and 11"
Solvent Constituents	-	-	M-R
2,3-Butanediol	$C_4H_{10}O_2$	D	L
Decane	$C_{10}H_{22}$	10	K
1-Octanol	$C_8H_{17}OH$	ol	I
1-Nonanal	$C_8H_{17}CHO$	al	G
2,6-Dimethyl Phenol	$(CH_3)_2C_6H_3OH$	P	F
Undecane	$C_{11}H_{24}$	11	H
2,6-Dimethyl Aniline	$(CH_3)_2C_6H_3NH_2$	A	E
C_{10}-Methylester	$C_{10}H_{19}O_2(CH_3)$	10e	D
C_{11}-Methylester	$C_{11}H_{21}O_2(CH_3)$	11e	C
Dicyclohexyl Amine	$C_{12}H_{23}N$	am	B
C_{12}-Methylester	$C_{12}H_{23}O_2(CH_3)$	12e	A

Certainly different selectivities and efficiencies are observed with the two chromatograms. The resolution is favorable in the temperature programmed elution for the volatile compounds, the contrary is true for the least volatile substances.

Effects of Film Thickness. "Figure 6" shows three chromatograms of the Grob sample mixture. All columns were 15 m long and coated with the stationary phase SE-54. The temperature gradient was $1^{\circ}C/min$ starting at $60^{\circ}C$. From top to bottom the column had a film thickness of 0.5, 1.0 and 2.0 µm, respectively.

A first inspection of the chromatograms reveals that the column with the thinnest film had the best overall resolution, demonstrated by the base line resolution of peaks F, G and H. A more detailed look shows that with increasing film thickness the efficiency increases, but at the same time the selectivity decreases. This is particularly so for the separation of peaks B and C. However, the resolution is increased for volatile substances when the film thickness is increased. With rather thick film columns it is possible to separate the main constituents of earth gas as is demonstrated in "Figure 7".

Effects of Column Length. The influence of column length on the resolution and efficiency is known. To increase the resolution of a homologous substance pair by two, we must increase the column length by a factor of four.

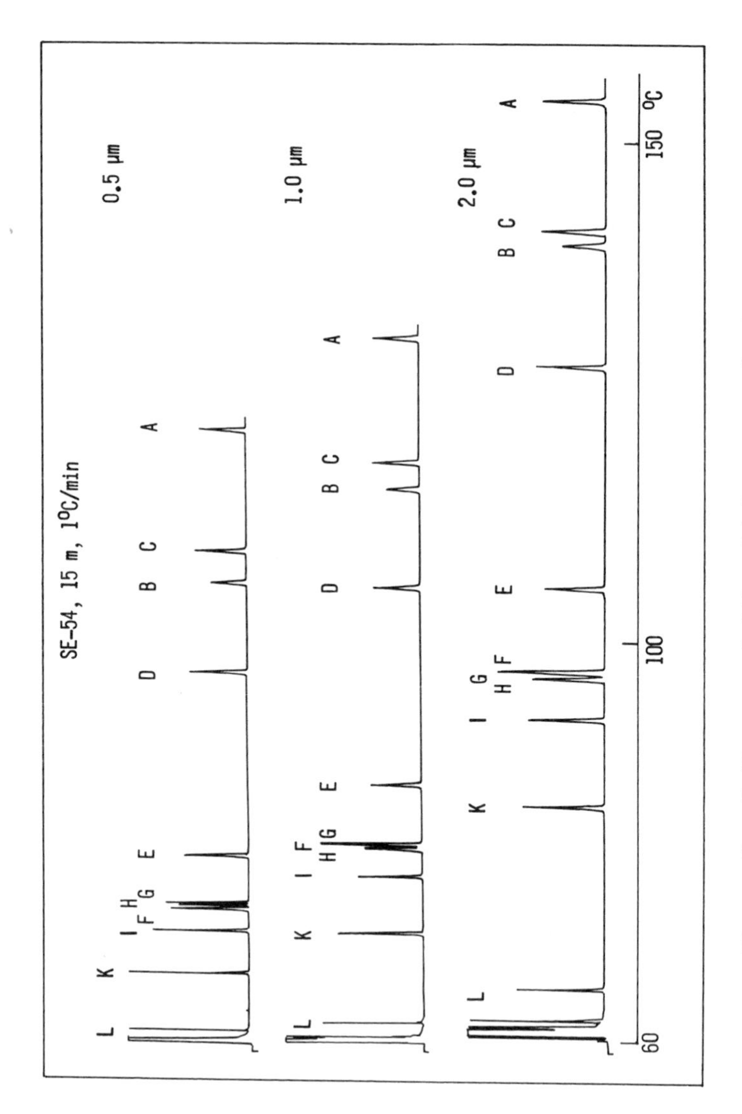

Figure 6. Influence of film thickness on elution temperature and chromatographic resolution.

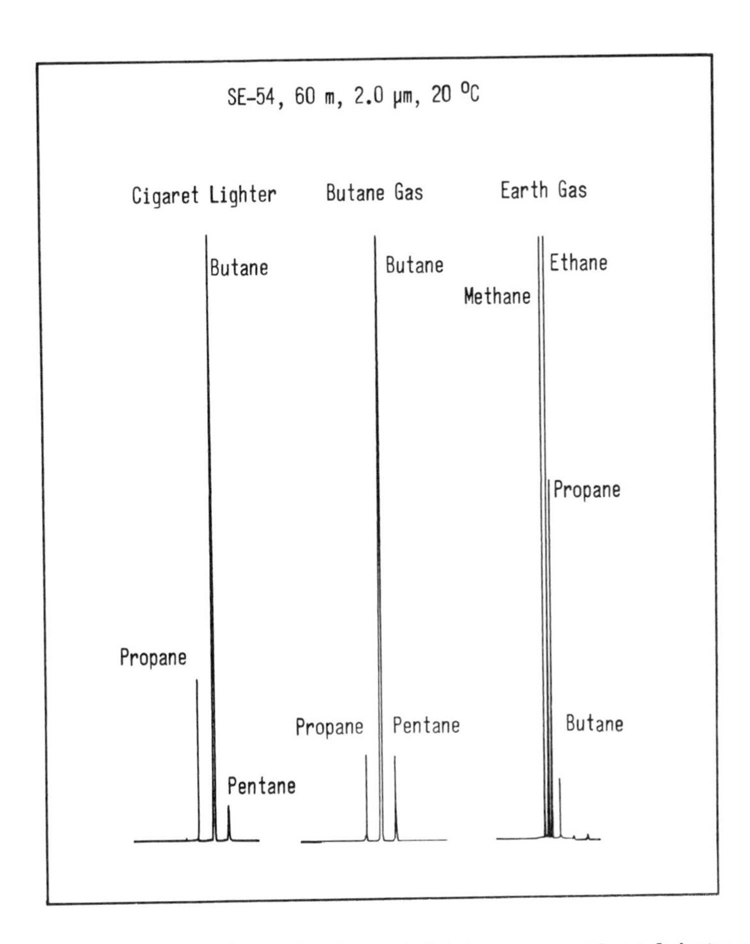

Figure 7. Separation of cigaret lighter, earth and butane gas with thick film capillary columns.

The situation may be quite different for non-homologous constituents."Figure 8" represents the chromatograms of SE-54 capillary colums all coated with a 0.5 μm film. The length of the columns were from top to bottom 15, 30 and 60 m, respectively. Considering the last two peaks A and C, one observes separation numbers of 27, 39 and 55 for the top, middle and bottom chromatogram, respectively, as expected from theory. Substrates A and C are indeed members of a homologous series.

But focussing on the performance of the 30 m column, one recognizes that peaks F, G and H are not resolved. Apparently, constituents F and G have similar retention times.

Increasing the column length to 60 m is not satisfactory either. However, one observes a true base-line resolution with the shortest column on top. A rather unexpected result from efficiency theory, but a common case in the industrial world. Obviously, substrates F and G have a different dependence of their relative retention times on a change of temperature (for explanation see also "Figure 12").

Peak Inversions. Peak inversions are quite common in capillary gas chromatography. In "Figure 9" peak inversion is achieved by selecting different temperature rates with the same column.

In "Figure 10" the inversion is observed based on columns with different film thickness. These examples demonstrate that in many cases a better chromatographic separation is obtained taking advantage of favorable temperature effects. These are best explained by a consideration of retention indices.

"Figure 11" represents the simultaneous effects of changing film thickness and column length. These examples also demonstrate that there are techniques available to produce capillary columns of high precision and reproducibility (8). The columns are characterized from top to bottom by a film thickness of 2.0, 1.0 and 0.5 μm and a column length of 15, 30 and 60 m, respectively.

Kovats Indices. The Kovats Index of a substance X, referring to a chromatogram run at temperature T and with a stationary phase P is calculated according to "Equation 8"

$$I_X^{T,P} = \frac{\log \frac{t'_{m_X}}{t'_{m_B}}}{\log \frac{t'_{m_A}}{t'_{m_B}}} (I_A^{T,P} - I_B^{T,P}) + I_B^{T,P} \qquad (8)$$

B and A refer to the alkanes eluting before and after substrate X.

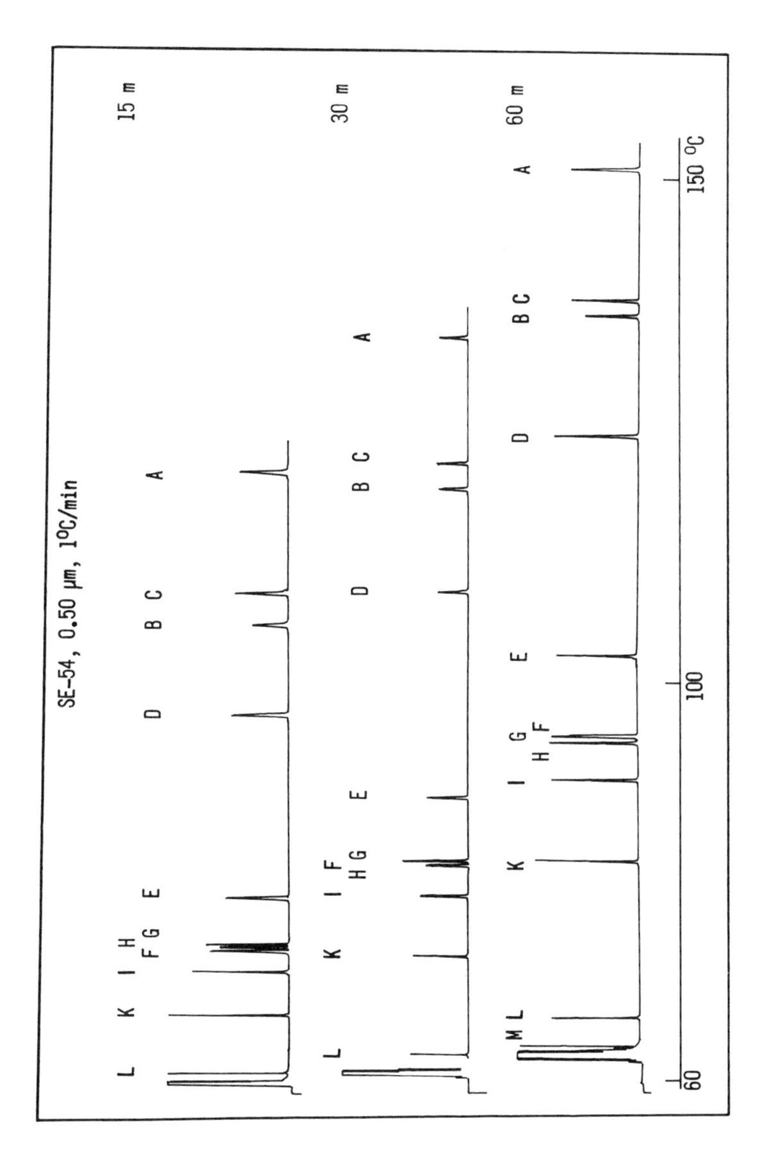

Figure 8. Influence of column length on elution temperature and chromatographic resolution.

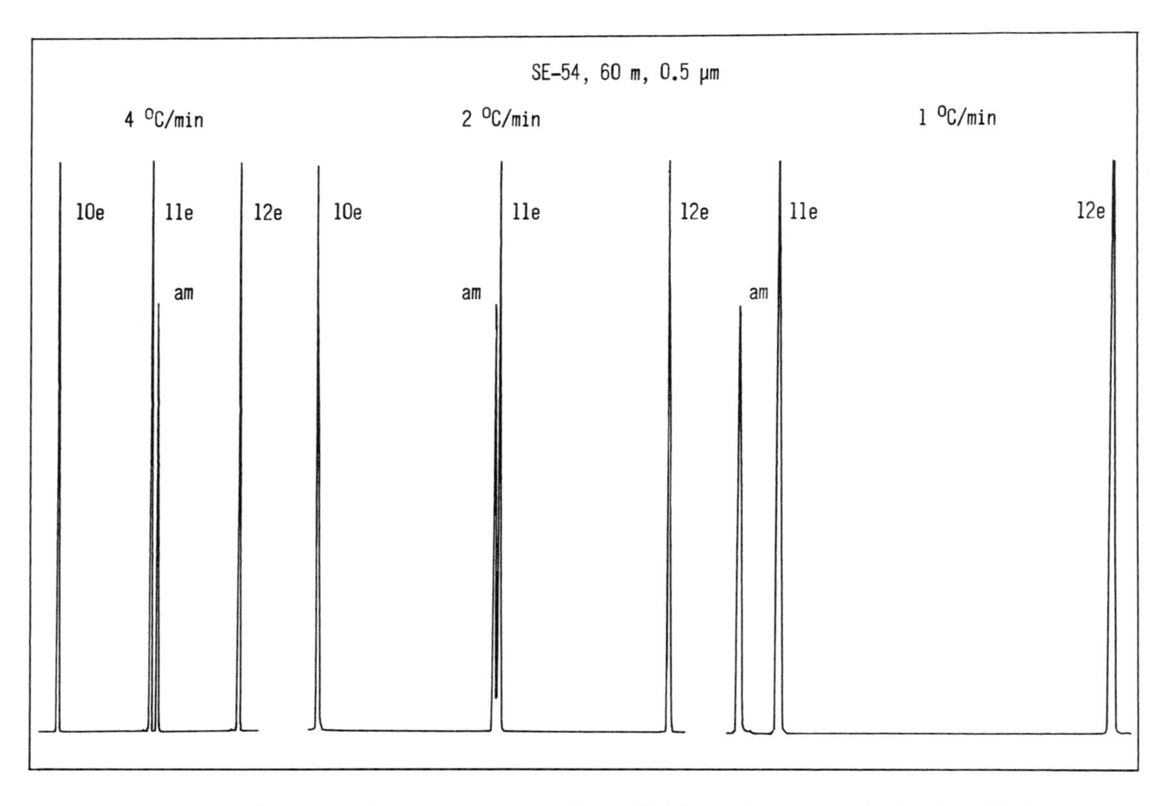

Figure 9. Peak inversions based on different temperature programs.

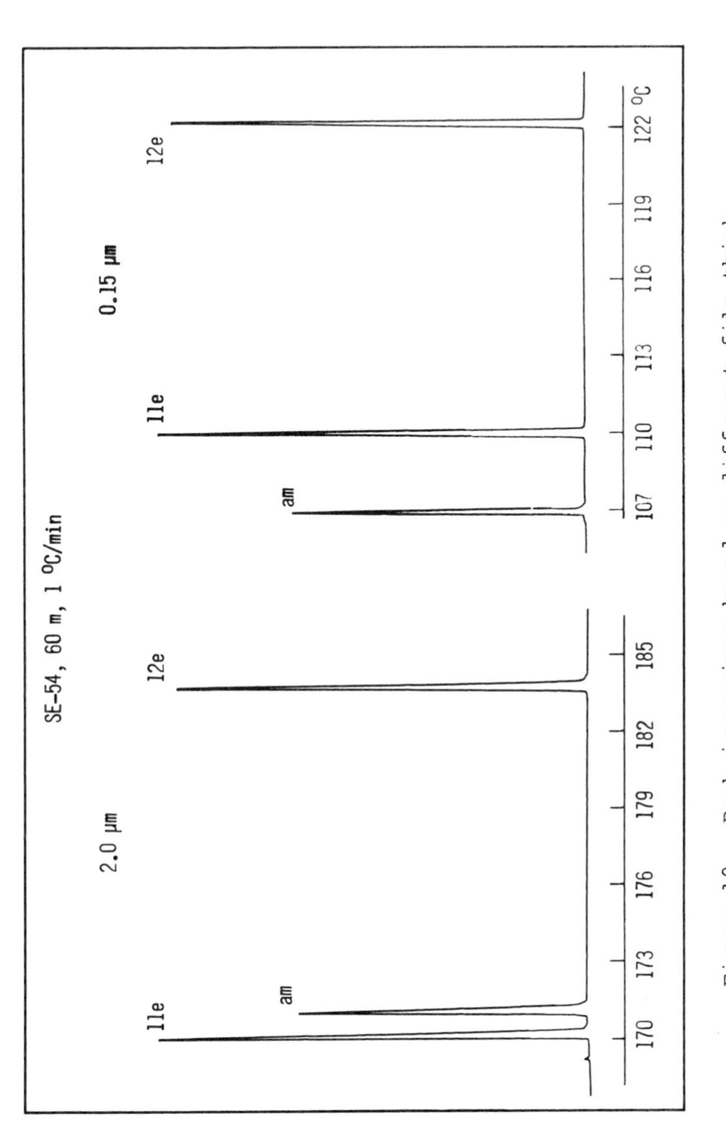

Figure 10. Peak inversions based on different film thickness.

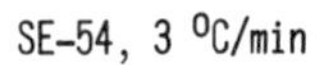

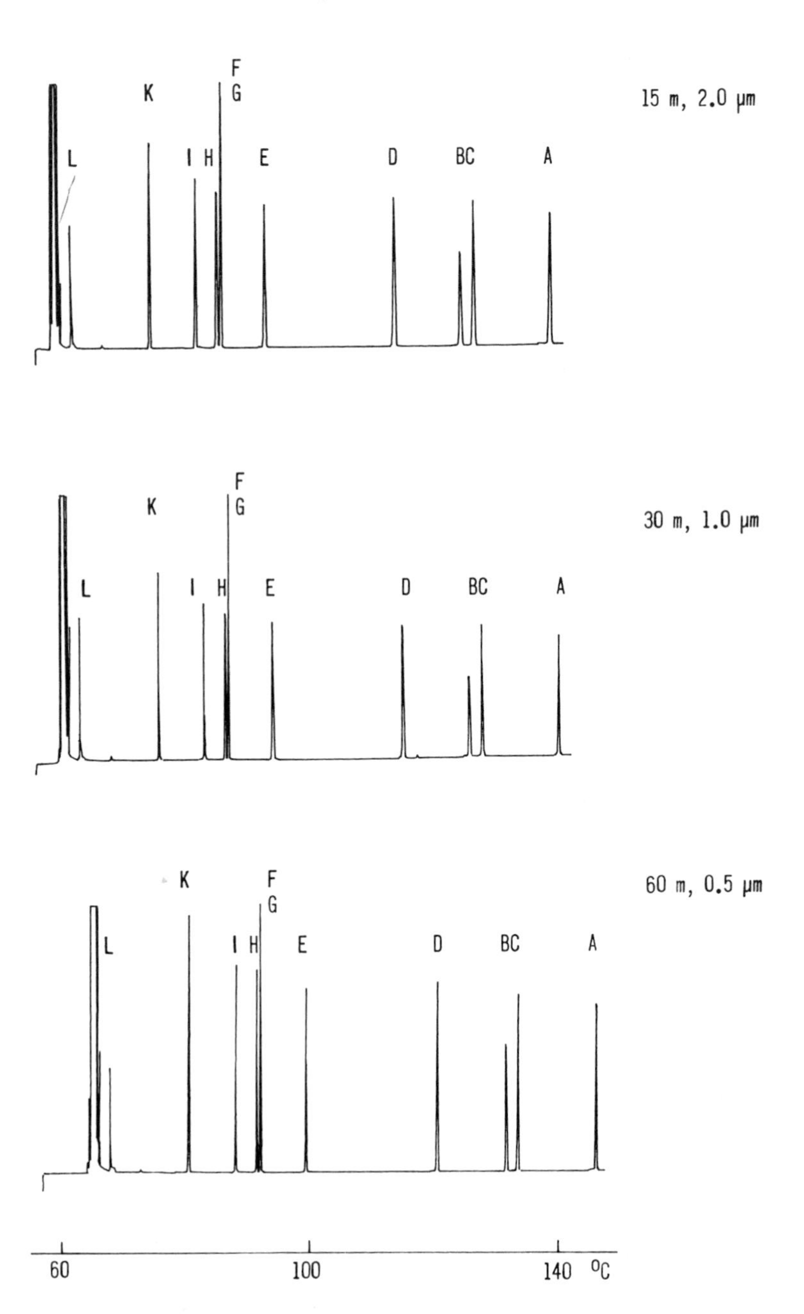

Figure 11. Simultaneous effects of various film thickness and column length.

"Figure 12" represents the temperature dependence of the Kovats indices for the substances F, G and H from the Grob Test Mixture. "Table V" summarizes the values of the prominent components of the Grob Test Mixture for OV-1, SE-54, OV-1701 and PEG 40M. With PEG 40 M one observes a negative $\Delta I/\Delta t$ value for alcohols.

We conclude that on a given capillary column we fail to separate two components when these exhibit the same Kovats index as well as identical $\Delta I/\Delta t$-terms. A careful inclusion of temperature effects and taking advantage of the substance dependent elution temperatures may be advantageous over other precautions, including fancy column switching techniques. A pragmatic approach, paired with appropriate experience and know-how may be superior over a restricted theoretical approach in many problem solving attempts.

Industrial Tendencies

Capillary GC on one side and ultrahigh resolution HPLC and SFC on the other need further development to meet the demands of today and the near future.

High resolution chromatography is not a simple matter of column performance alone, but includes injection and detection techniques as well. There is a definite tendency in industry to search for improvements of the situation in ultrahigh resolution chromatography. They may be summarized as follows:

New Separation Techniques

- Capillary GC
  - immobilized phases (there is a trend for immobilization of polar separation phases)
  - separation phases with specific effects (e.g. optical isomer separation)
  - application of small bore capillaries
  - column switching and heart cutting
- Liquid Chromatography
  - reversed phase and size exclusion systems
  - ion chromatography
  - ion pair chromatography
  - microbore HPLC
  - column switching
  - transfer through permeation membrane techniques
  - field flow fractionation
- Supercritical Fluid Chromatography (SFC)

Improved Injection Systems

- automated on-column injection
- trapping techniques
- enrichment techniques
- microwave injectors

Table V. Kovats Indices and $\Delta I/\Delta t$-Values at 150°C

Compound	OV-1		SE-54		OV-1701		PEG 40M	
	I	$\Delta I/\Delta t$ x 10	I	$\Delta I/\Delta t$ x 10	I	$\Delta I/\Delta t$ x 10	I	$\Delta I/\Delta t$ x 10
1-Octanol	1051.7	0.1	1069.3	0.1	1181.4	0.0	1544.1	-1.8
1-Nonanal	1088.6	1.2	1110.0	1.0	1209.6	2.2	1411.7	2.2
2,6-Dimethyl Phenol	1095.3	3.3	1126.4	3.6	1294.4	3.7	1888.4	4.3
2,6-Dimethyl Aniline	1157.9	4.8	1194.7	4.8	1340.3	5.7	1840.6	7.7
10-Methylester	1305.3	0.4	1324.1	0.0	1399.5	0.6	1603.3	0.8
11-Methylester	1406.6	0.1	1424.3	0.1	1499.9	0.7	1704.5	1.1
Dicyclohexyl Amine	1415.8	6.5	1434.7	6.9	1486.3	7.6	1669.2	9.3
12-Methylester	1507.2	0.1	1523.1	0.3	1600.3	0.8	1805.9	1.0

Improved Detector Systems

- tandem detectors such as ECD/FID in GC
- electrochemical detectors in HPLC
- optical sensors (optrodes)
- chemical field effect transducer (chemfet)
- laser-based detectors (e.g. photoacoustic detectors, laser-induced fluorescence, etc.)

Hyphenated Systems

- GC/TEA
- HPLC/TEA
- HPLC/MS
- SFC/MS
- SFC/FT-IR

Interdisciplinary Influence from other Analytical Techniques

- flow injection analysis
- electrochemical development
- laser technology
- biochemical and enzyme immobilization techniques

These tendencies are illustrated with an example from microbore column LC/MS. "Figure 13" describes the LC/MS interface used and gives a schematic view of the interface principle. "Figure 14" represents the mass spectrum of an analyzed sample, involving a compound with a molecular weight of 530.

Conclusions

The state of the art in ultrahigh resolution chromatography allows the industrial analyst already now to introduce this technique in his daily work. However, further improvements are needed to make full use of the inherent potential in industrial applications.

Legend of Symbols

E	Chromatographic efficiency
E_A, E_B	Efficiency of peak A and B, respectively
H	Height equivalent to a theoretical plate (HETP)
H_{Min}	Minimum HETP
H_{eff}	Effective HETP
I, $I_X^{T,P}$	Kovats index of substrate X at temperature T and refering to the separation phase P
k', k'_B	Capacity factor
N	Number of theoretical plates for column chromatography
N_{Max}, N_{eff}	Maximum and effective N, respectively

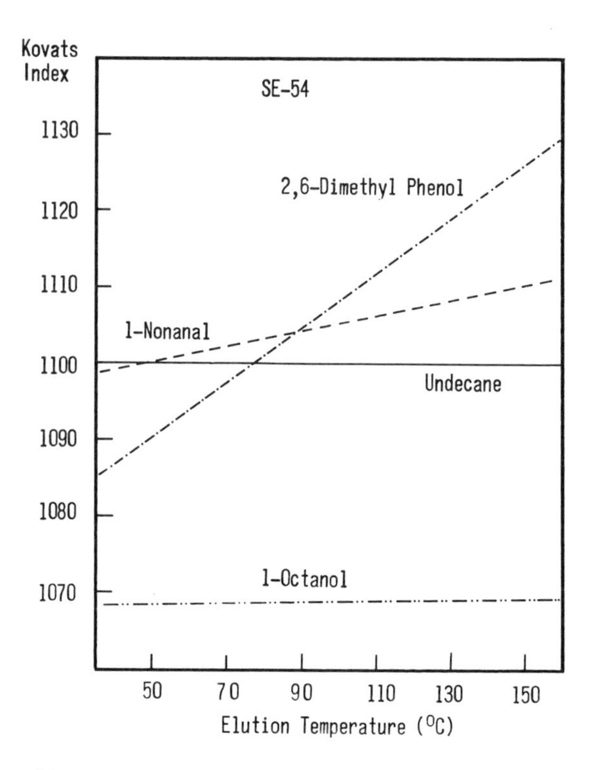

Figure 12. Temperature dependence of Kovats indices for 2,6-dimethyl phenol, 1-nonanal, undecane and 1-octanol.

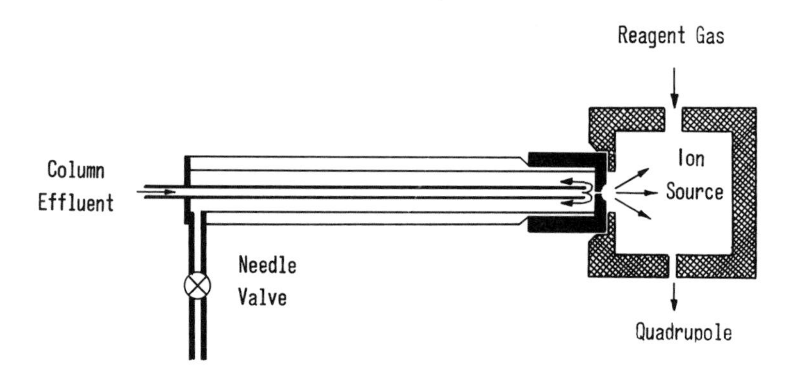

Figure 13. Schematic of microbore LC/MS interface.

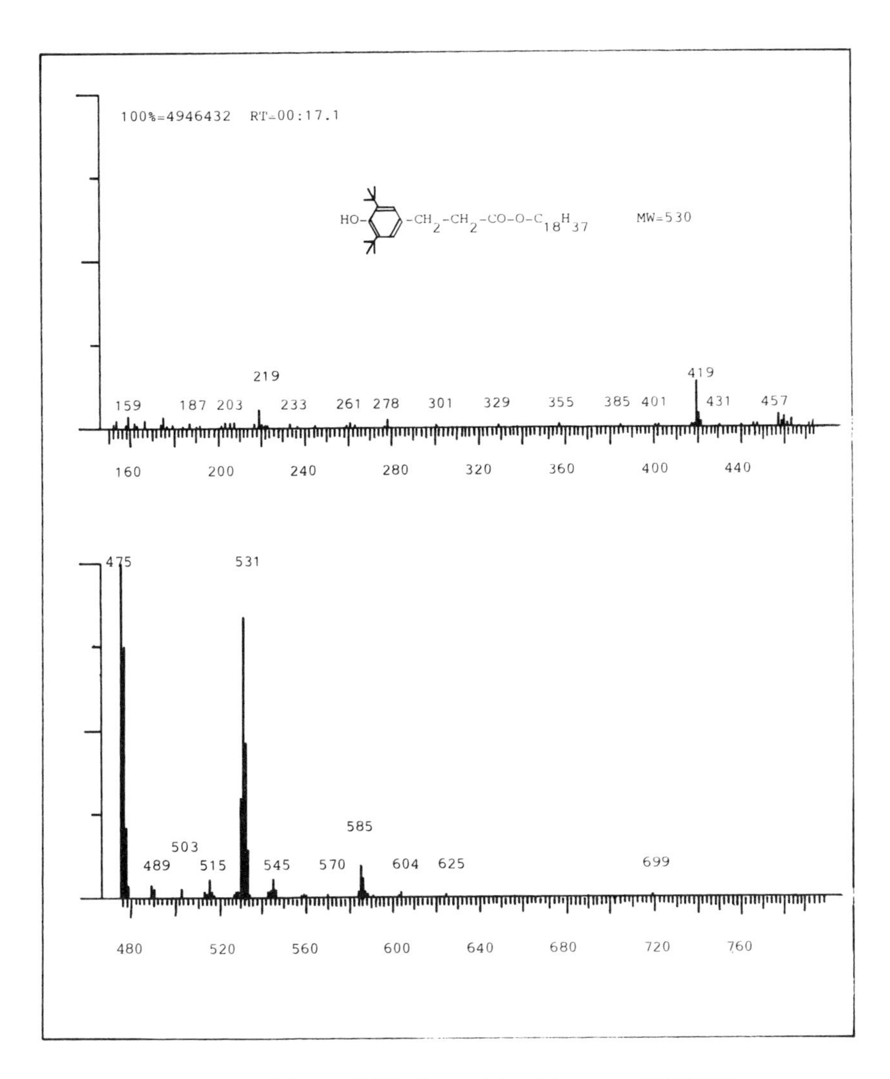

Figure 14. LC/MS investigation on IRGANOX.

R_s	Chromatographic resolution
$S_{A,B}$	Selectivity of separation bewteen peak A and B
t'_{m_X}	Corrected retention time used to calculate Kovats index
t_{R_A}, t_{R_B}	Retention time of substrate A and B, respectively
$\bar{u}$	Mean linear velocity of the mobile phase
W_A, W_B	Peak width of substrate A and B, respectively
α	Separation factor
ECD	Electron capture detector
FID	Flame ionisation detector
FT-IR	Fourier transform infra red
GC	Gas chromatography
HPLC	High performance liquid chromatography
LC	Liquid chromatography
MAK	Maximale Arbeitsplatz Konzentration
MS	Mass spectrometry
SFC	Supercritical fluid chromatography
TEA	Thermal energy analyzer
TLV-TWA	Threshold limit value-time weighted average

Literature Cited

1. Widmer, H. M.; Grass, G. Proc. 3rd Intern. Symp. Loss Prev. Safety Prom. Process Ind. 1980, Vol II, p. 3/268.
2. Grass, G.; Widmer, H. M. Swiss Chem 1981, 3, 117.
3. Golay, M. J. E. "Gas Chromatography"; Desty, D. H., Ed.; Butterworths London: 1958, p. 36.
4. Golay, M. J. E. Nature 1963, 199, 370.
5. Kaiser, R. "Chromatographie in der Gasphase. II. Kapillar-Chromatographie, Dünnfilm- und Dünnschichtkapillar-GC"; Biographisches Institut: Mannheim/Wien/Zürich, 1975; p. 71.
6. Van Deemter, J. J.; Zuiderweg, F. J.; Klinkenberg, A. Chem. Eng. Sci. 1956, 5, 271.
7. Grob, K.; Grob, G.; Grob, K. J. Chrom. 1981, 219, 13.
8. Widmer, H. M.; Grolimund, K. Proc. 5th Intern. Symp. Capil. GC Hindelang 1981, p. 751.

RECEIVED January 25, 1984

Author Index

Subject Index

L

M

N

Production by Anne Riesberg
Indexing by Susan Robinson
Jacket design by Anne G. Bigler

Elements typeset by Hot Type Ltd., Washington, D.C.
Printed and bound by Maple Press Co., York, Pa.

RECENT ACS BOOKS

"Chemistry of Combustion Processes"
Edited by Thompson M. Sloane
ACS SYMPOSIUM SERIES 249; 286 pp.; ISBN 0-8412-0834-4

"Geochemical Behavior of Disposed Radioactive Waste"
Edited by G. Scott Barney, James D. Navratil, and W. W. Schulz
ACS SYMPOSIUM SERIES 248; 470 pp.; ISBN 0-8412-0831-X

"NMR and Macromolecules:
Sequence, Dynamic, and Domain Structure"
Edited by James C. Randall
ACS SYMPOSIUM SERIES 247; 282 pp.; ISBN 0-8412-0829-8

"Geochemical Behavior of Disposed Radioactive Waste"
Edited by G. Scott Barney, James D. Navratil, and W. W. Schulz
ACS SYMPOSIUM SERIES 246; 413 pp.; ISBN 0-8412-0827-1

"Size Exclusion Chromatography: Methodology and
Characterization of Polymers and Related Materials"
Edited by Theodore Provder
ACS SYMPOSIUM SERIES 245; 392 pp.; ISBN 0-8412-0826-3

"Industrial-Academic Interfacing"
Edited by Dennis J. Runser
ACS SYMPOSIUM SERIES 244; 176 pp.; ISBN 0-8412-0825-5

"Characterization of Highly Cross-linked Polymers"
Edited by S. S. Labana and Ray A. Dickie
ACS SYMPOSIUM SERIES 243; 324 pp.; ISBN 0-8412-0824-9

"Polymers in Electronics"
Edited by Theodore Davidson
ACS SYMPOSIUM SERIES 242; 584 pp.; ISBN 0-8412-0823-9

"Radionuclide Generators: New Systems
for Nuclear Medicine Applications"
Edited by F. F. Knapp, Jr., and Thomas A. Butler
ACS SYMPOSIUM SERIES 241; 240 pp.; ISBN 0-8412-0822-0

"Polymer Adsorption and Dispersion Stability"
Edited by E. D. Goddard and B. Vincent
ACS SYMPOSIUM SERIES 240; 478 PP.; ISBN 0-8412-0820-4

"Archaeological Chemistry--III"
Edited by Joseph B. Lambert
ADVANCES IN CHEMISTRY SERIES 205; 324 pp.; ISBN 0-8412-0767-4

"Molecular-Based Study of Fluids"
Edited by J. M. Haile and G. A. Mansoori
ADVANCES IN CHEMISTRY SERIES 204; 524 pp.; ISBN 0-8412-0720-8